Francisco Tomaz Pacífico Júnior

Environmental Responsibility in the Hotel Sector

Francisco Tomaz Pacífico Júnior

Environmental Responsibility in the Hotel Sector

A study carried out in the city of Mossoró/RN

ScienciaScripts

Imprint

Any brand names and product names mentioned in this book are subject to trademark, brand or patent protection and are trademarks or registered trademarks of their respective holders. The use of brand names, product names, common names, trade names, product descriptions etc. even without a particular marking in this work is in no way to be construed to mean that such names may be regarded as unrestricted in respect of trademark and brand protection legislation and could thus be used by anyone.

Cover image: www.ingimage.com

This book is a translation from the original published under ISBN 978-620-2-04930-6.

Publisher:
Sciencia Scripts
is a trademark of
Dodo Books Indian Ocean Ltd. and OmniScriptum S.R.L publishing group

120 High Road, East Finchley, London, N2 9ED, United Kingdom
Str. Armeneasca 28/1, office 1, Chisinau MD-2012, Republic of Moldova, Europe
Printed at: see last page
ISBN: 978-620-7-24427-0

Copyright © Francisco Tomaz Pacífico Júnior
Copyright © 2024 Dodo Books Indian Ocean Ltd. and OmniScriptum S.R.L publishing group

SUMMARY

ACKNOWLEDGMENTS 2
SUMMARY 4
1   INTRODUCTION 5
2   THEORETICAL FRAMEWORK 13
3   METHODOLOGY 40
4   ANALYSIS AND DISCUSSION OF RESULTS 47
5   FINAL CONSIDERATIONS 60
6   REFERENCES 63
7   APPENDICES 71

To my parents, Francisco Tomaz Pacifico (*in memorian*) and Francisca Ilma.

ACKNOWLEDGMENTS

Throughout our lives, no matter what we do, we certainly couldn't have done it without the effective participation of a significant number of people. The act of remembering these people at this time is a way of expressing my gratitude and thanking them for all the help they gave me during the production of this research and the completion of this dissertation. So, in this simple but sincere gesture, I would like to say thank you:

I would like to thank God for the opportunities I have been given in life, above all for having protected me during these more than 24 kilometers of travel and journeys during the completion of this master's degree, for having met people who have only made me greater as a human being, but I would also like to thank him for having lived through difficult moments, situations that have been fundamental as an example of learning and a reason to test my faith.

To my parents, especially my mother Francisca Ilma, because I don't know what I would have been without her education and for being a living example of a human being, of character and honesty. You have always made me believe that I can be the best. I love you unconditionally, my mother.

I would like to thank my advisor, Professor Dr. Lais Karla da Silva Barreto, who believed in my potential in a way that even I didn't think I could match. She was always available and willing to help, wanting me to take advantage of every second during the research to absorb some kind of knowledge. She made me realize that there is much more than researchers and consequences behind a dissertation. In short, she was certainly a fundamental part of the outcome of my scientific work.

I would also like to thank all my friends, because I can say with conviction that I have the best ones by my side, and that they have helped me in every way possible. I would especially like to thank Pablo Marlon and Alamo Duarte, who are also master's students at this institution.

I would like to thank my master's colleagues, who were always very helpful in carrying out group activities.

I would like to thank the hotel establishments surveyed for opening their doors so that I could carry out my research.

I would like to thank Universidade Potiguar - UnP for the opportunity it gave me, for giving me the chance to fulfill my dream of a master's degree. It gave me more than just the search for technical and scientific knowledge, but a lesson in life.

Finally, I would also like to thank those who have contributed in some way to this work. People who believed in my potential, encouraged me and participated with me, directly and indirectly, in this journey.

"A mind that is open to a new idea never returns to its original size"

(Albert Einstein).

SUMMARY

The interdisciplinary field of Management keeps pace with the innovations of organizations, discussing, among other things, sustainable development and environmental responsibility as a strategic focus. Directly linked to and totally dependent on the environment, tourism is one of the fastest growing industries in the world. Responsible for 9% of international GDP, tourism is one of the main economies on the world stage. As part of the tourist industry, the hotel sector is directly linked to current environmental problems and has a significant influence on environmental impacts, since many hotel developments are located in natural areas, historic towns and even in regions protected by environmental legislation. Given this context, this research aims to understand the perception of hotel chain managers in Mossoró, Rio Grande do Norte, about Corporate Environmental Responsibility and sustainability practices. The research is based on a case study of the four main hotel organizations in the city, with a descriptive nature and qualitative approach, using a semi-structured interview script adapted from Santos (2004) as a data collection instrument, applied to managers responsible for decisions regarding environmental practices and sustainability, and then submitted to content analysis. The results showed a lack of formal environmental practices in the hotels studied. With regard to the advantages of implementing environmental practices, it was found that managers' attention to their development is more focused on reducing costs and efficiency. As for the barriers, the difficulty in raising awareness among hotel staff and resistance to change were identified as the main problems faced. Several environmental education actions were highlighted by those surveyed during data collection, among which we can highlight the proposal for guests to reuse their towels in order to increase water savings from washing them; educational talks for employees, as well as measures to reduce water and energy consumption, and sewage treatment and solid waste separation. There is a need for further research into environmental responsibility in the city's businesses, with the aim of raising awareness among entrepreneurs and customers of the need to acquire and maintain a sound and permanent social and environmental attitude.

Keywords: Environmental responsibility. Tourism. Hospitality.

1 INTRODUCTION

The introduction discusses the topic of environmental issues in today's tourism and hospitality organizations and how they have been concerned with the challenges of preserving natural resources and the environment, as well as seeking a competitive edge in an increasingly competitive market. Next, the problematization, objectives, justification and structure of the work are presented.

1.1 CONTEXT

Identifying future trends and anticipating market changes have become determining factors for competitiveness among organizations. These factors are decisive and fundamental for dealing with uncertainty, adapting creatively and quickly to important changes, taking advantage of unexpected opportunities and thus managing to stay ahead in an increasingly flexible and dynamic economy.

The disciplinary field of Administration keeps pace with the innovations of organizations, discussing, among other things, sustainable development and environmental responsibility as a strategic focus.

The environment, as well as issues related to it, are the target of several authors from various fields of knowledge. In these approaches, complexity, a systemic view, recursion and interdisciplinarity are presuppositions for the new world view that aims for Sustainable Development - SD (GIESTA, 2013).

The implications of the ecological imbalance caused by industrialization are increasing all the time, with consequences for the entire production chain, from the extraction of raw materials and the manufacture of products to their distribution through pipelines. The impact of the ever-increasing damage caused to the environment has led to a gradual increase in global environmental awareness, with direct repercussions on companies' operating procedures and the conduct of their business (CAVALCANTI, 2006).

In this way, it is possible to observe that the environment has been recognized over the years not only as a source of resources, but also as a good to be preserved.

Organizations are becoming increasingly aware of environmental concerns, due to a number of factors, but above all due to customers taking an increasingly rigid stance towards companies that have a good image in the market, in other words, companies that adopt sustainable measures in their processes and show an interest in minimizing the damage caused to the environment, both in the manufacture of their products and in the performance of their services.

Environmental issues and the scarcity of natural resources have long been a cause for concern in organizations around the world. For this reason, companies in all sectors of the economy are paying significant attention to the environment, especially in the tourism and hospitality sectors, which are mainly dependent on the natural environment for their survival. Faced with the need for this new attitude, organizations are interested in actions that reduce environmental aggression. This raises the challenge of reconciling economic growth and social development in a sustainable way with the environment and its natural resources, following the criteria required by environmental standards and legislation.

As it is an activity that depends almost exclusively on natural resources for its existence, tourism requires sustainable planning methods and environmental management systems, which are vital for the harmonious development of tourism (BORGES, 2011).

Tourism is one of the fastest growing industries in the world. According to the World Tourism Organization (UNWTO, 2015), in 2014 the number of people who practiced tourism exceeded one billion, in addition to having moved the international economy by US $ 1.5 trillion for the first time, a fact that reinforces the idea of the segment's growth.

With regard to the impact of tourism on the Brazilian economy, the National Tourism Plan (PNT) (2015) reveals that Brazil's tourism GDP is ranked sixth in the world by the *World Trade Tourism Council* (WTTC) (2015), behind countries such as the United States, China, Japan, France and Italy. In addition to its privileged position in the ranking, the Brazilian government is hoping to reach third place after major events in the country, such as the World Cup in 2014 and the Olympics in 2016. According to the forecast in the WTTC report (2015), Brazil should overtake France by 2022 in terms of the impact of tourism on GDP, but the Brazilian government is forecasting third place, ahead of Japan.

Tourism already accounts for 3.7% of Brazil's Gross Domestic Product (GDP). From 2003 to 2009, the sector grew by 32.4% while the Brazilian economy expanded by 24.6% (MTUR, 2014). However, there is a need for a unity of effort between all the agents involved in the tourism sector in order to guarantee the position that the country really deserves, especially when considering factors such as: its continental size; its geographical location; its rich natural, cultural and historical heritage, added to its rich biodiversity.

According to Abreu (2001), although tourism in Brazil may have already reached a consolidated level in national economic policy, much still needs to be done to improve our country's position in this competitive ranking, preparing it to offer quality services to an increasingly demanding class of clients.

From the above information, it can be concluded that the tourism sector has a significant position in the economy of several countries, and this activity should be given due attention, as it is a sector of the economy, which like so many others, be they primary, secondary or tertiary, has defenders and critics.

With regard to the authors who believe in the benefits of the tourism sector, Medeiros and Morais (2013) provide the following information: since the beginning of tourism, the tourism sector has been considered harmless to the environment, especially when compared to other sectors of industry, which, for example, need to extract natural resources in order to produce their products or offer their services.

However, other authors (CRUZ, 2001; Barbieri 2007; CAON, 2008; TUNG & AYCAN, 2008, among others) believe that if tourism is carried out in an unplanned way, even if it doesn't require the actual extraction of natural resources, this segment needs these resources to exist and, if the activity is carried out in an unsustainable way, it can cause irreversible impacts on the environment.

Cruz (2001) presents some of the impacts that tourism activities carried out in a disorderly manner can cause:

- Increased generation of solid waste;
- Increased demand for electricity;
- Increased vehicle traffic, resulting in reduced air quality;
- Siltation of the coast, due to human actions, with the destruction of corals;
- Contamination of river and sea water due to the increase in untreated sewage;
- Changes in the lifestyle of native populations;
- Degradation of the landscape, due to inadequate building construction; among others.

On the other hand, with careful planning and good practices, tourism can contribute to sustainable development. A growing body of literature has documented efforts made by tourism companies to reduce pollution and improve the sustainability of their businesses, including zero waste initiatives as one of their practices (PANATE, 2015).

Tourism has been identified as a driving force for local community development and poverty reduction in less developed countries (IVANOV, 2012).

As well as developing the local community, another aspect of tourism deserves dedicated attention: the environmental issue, since it is a finite resource and in the search for excellence in the services offered, it is gaining expression through increasingly rigorous clients seeking services that take care of the environment.

As part of the tourism industry, the hotel sector is directly linked to current environmental problems and has a significant influence on environmental impacts, since many hotel developments are located in natural areas, historic towns and even in regions protected by environmental legislation (CAON, 2008).

The hotel industry has become increasingly diversified, more aggressive and in some properties, whether large hotel chains or family businesses, management has acquired less traditional habits and forms, emphasizing environmental issues (VARUM et al, 2011).

The hotel sector in Brazil has undergone a major conceptual change in the last ten years. One of the most important actions included in the hotel classification matrix is Environmental Management.

The hotel, as well as other productive and service-providing activities, occupy space in a certain environment, which will include physical and operational facilities that will generate waste, causing environmental impacts, degrading this environment in some way, and depending on the concerns during project design, construction and operation, these impacts can have varying degrees of aggression, and can be: permanent, frequent, sporadic and rare. Depending on the case, remediation or recovery of this environment may become irreversible (ALVES, 2012).

By implementing and executing conscious and responsible environmental management practices, the organization will be able to minimize not only the direct environmental risks, but also the risks related to the institutional image of the institution (VALLE, 1995).

Today, not only the large hotel companies, but also the small businesses that are part of the tourist trade are showing concern for the environment through their operating procedures. These procedures are carried out through the implementation of strategic management - EMS - in their production chains, and according to the size of the expected impact, they develop the Operational Manual of the environmental management system, with all the procedures and resources to plan, implement and maintain the environmental policy.

Taking the above information as a reference, it can be assumed that hotel companies should implement measures in their operational routines to develop sustainable procedures, starting up environmental management systems in their establishments to help them comply with environmental regulations and adapting their activities so that they can comply with environmental certifications, thus acquiring a competitive edge over their competitors and, as a result, taking the necessary measures to protect the environment.

Although tourism is considered to be one of the main activities responsible for the socio-economic and cultural growth of a region, this segment is a recent field of study within the Human Sciences. It is also worth noting that this activity has several conditions inherent in its awareness of the environment.

Even if tourism is not directly linked to environmental damage, the consequences of poor planning can lead to irreversible environmental problems. As Barbieri (2007) points out, solving environmental problems or minimizing them requires organizational changes and a restructuring in the awareness of managers, who must begin to consider the environment in their decisions and adopt administrative, social and technological procedures that contribute to maximizing the environment's carrying capacity.

1.2 PROBLEMATIZATION

Taking the above information as a yardstick and bearing in mind that environmental issues are increasingly in evidence these days, whether for reasons related to the pressure exerted by governments on organizations, pressure exerted in various ways, whether through the creation of laws or the adoption of measures to show concern for the environment, or the adoption of measures to ensure that companies adopt a stance that shows concern for environmental preservation, or even because of society's awareness of the emergence of customers who are increasingly looking for environmentally friendly products and services.

In view of the above, the central problem of this research is the following question: **what is the perception of managers of hotel organizations in Mossoró, Rio Grande do Norte, about Corporate Environmental Responsibility and sustainability practices?**

1.3 OBJECTIVES

1.3.1 General

To understand the perception of the managers of the hotel chain in Mossoró, Rio Grande do Norte, on the practices of Corporate Environmental Responsibility and sustainability.

1.3.2 Specifics

- To analyze and describe the environmental management and sustainability practices implemented by managers in the main hotel enterprises in the municipality of Mossoró, Rio Grande do Norte;
- To ascertain the advantages and challenges of implementing environmental practices from the point of view of managers;
- To learn about environmental education actions and how they are transmitted to the professionals and clients of the organizations studied.

1.4 BACKGROUND

Mossoró is the second most populous municipality in the state of Rio Grande do Norte, strategically located between two important northeastern capitals, Natal and Fortaleza, 260 kilometers from the latter and 275 km from the capital Natal. It has a high level of foot traffic every day of the week, which ensures that the town's lodging establishments are busy every month of the year and not just during the high season. This region is located in the center-west of the state and has a flat geography. Its name is a reference to the local landscapes, with the predominance of dunes and salt marshes with huge white hills (PORTAL, 2015).

The municipality of Mossoró is considered a hub city, as it serves as a reference and support for several surrounding municipalities. In 2014, it had a population of approximately 284,288 inhabitants, according to IBGE data (2014). This places it as the 20th largest city in the Northeast, in a transitional region between the coast and the hinterland, 42 kilometers from the coast.

In its economy, salt mining has always been a major activity. However, with the subsequent exploitation of oil, the city has experienced great economic and social development.

Until the mid-1980s, Mossoró's economy was mainly sustained by the local salt industry, which to this day still supplies 60% of the product consumed in the country. At that time, melon plantations were introduced, which today employ more than sixty thousand people. In the following decade, oil royalties changed the city's economy (COUTINHO, 2010).

These activities also boost tourism in the city, especially the business tourism segment. Continuing on the subject of tourism, and in parallel with economic growth, the tourism industry in Mossoró has shown visible growth in recent years, especially in the area of cultural tourism.

The three great theatrical shows performed in the municipality allow visitors to experience the city's history. They are Chuva de Bala no Pais de Mossoró, which tells the story of Lampiao's defeat when he invaded the city; Auto da Liberdade, which presents the abolition of slavery and when Mossoró freed its slaves before the Aurea Law; and Oratòrio de Santa Luzia, which tells the story of the city's patron saint.

In addition to these cultural events, Mossoró also stands out for the Mossoró Cidade Junina, which is a major event held throughout the month of June and alludes to the celebrations of the June festivities. This event has earned the title of the city's main cultural attraction, capable of mobilizing a large part of the population and also attracting a significant number of tourists (MOSSORÓ, 2015).

Although the municipality doesn't have many tourist attractions with beautiful natural landscapes. Mossoró has strong tourist potential in the area of business tourism. As already mentioned, it has a privileged location and in terms of economic activities, it receives sales representatives and skilled labor from all over the country. This means that almost 100% of the city's hotels are occupied all year round.

In parallel with the growth in demand for what is considered business tourism, and taking into account the current concern of organizations with the preservation of the environment, hotels in various countries are also introducing environmental management into their daily activities, as they depend on natural resources that are under threat in order to continue their activities.

The importance of this study lies in the fact that discussions about concern for environmental preservation are increasingly present in everyday life, as well as the growing pressure exerted by governments in relation to the implementation of environmental practices in companies of various sizes and segments around the world.

For organizations, the issue is making it possible for managers to become more aware of the corporate environment, which has intensified since the emergence of standards, such as ISO 14000, for example, which is a series of standards developed by the *International Organization for Standardization* (ISO) and which establish guidelines for environmental management within companies. In Brazil, the first certified company was Bahia Sul Celulose S.A. in 1996. Here, certification is maintained by the Brazilian Association of Technical Standards (ABNT), and was therefore called ABNT NBR ISO 14001.

The research is also relevant to science, bringing new empirical contributions on the subject; and it is for the author, as it allows him to understand the concepts and benefits of environmental responsibility and to act as a defender of environmental causes, using the criterion of responsibility towards natural resources as indispensable for companies to acquire products or services.

1.5 WORK STRUCTURE

This research is structured in five sections, divided as follows: introduction; theoretical framework; methodological procedures; presentation and analysis of the results and finally, the final considerations.

The first section refers to the introduction, which presents the contextualization of the content of this work, as well as its problematic, the proposed objectives (general and specific), in addition to the justification, in which the reason for choosing the topic involved is explained.

The second section contains the theoretical framework, which sets out themes related to tourism and the environment, such as: sustainability; the origins of sustainable development; the relationship between tourism and sustainability; environmental management in hotels; corporate environmental responsibility; environmental management systems, among others.

The third section of this study presents the methodological procedures, which cover the type of research, the characterization of the research environment and subjects, the collection, processing and analysis of the research data (how the data was collected, through what procedures and how it was analyzed).

The fourth section presents and analyzes the results obtained from the research, based on the proposed objectives and the study and discussions based on the theoretical framework.

Finally, the fifth and last part of the paper presents the final considerations of the study, reviewing the objectives and comparing them with the final results, as well as presenting the limitations of the research and offering suggestions for future research on the subject of environmental responsibility in the hotel sector.

2 THEORETICAL FRAMEWORK

The main objective of this study is to identify the perceptions of managers in the Mossoró (RN) hotel chain when faced with a scenario of Corporate Environmental Responsibility and sustainability practices. But in order to be able to discuss managers' perceptions of the environmental management techniques used in certain hotel enterprises, it is necessary to provide the reader with a theoretical background on tourism (since the research is carried out in hotels), sustainability and Corporate Environmental Responsibility.

Therefore, the purpose of the first part of the framework is to present concepts of sustainable development from the point of view of various authors. It also mentions the applicability of sustainability to tourism and hospitality. In addition to addressing the concept of tourism from the outset of the activity, and to focus on the term: sustainable tourism, as well as the management practices carried out by hotel enterprises and, consequently, by tourism.

2.1 SUSTAINABILITY APPLIED TO TOURISM AND HOSPITALITY

Tourism activities have been developing on an increasing scale worldwide and the demand from developing countries such as Brazil, for example, is growing. Mainly due to its growth potential and the fact that tourism is a product that can only be consumed *on site*, this segment of the economy is taking on an important role as a local development strategy (IVARS, 2003).

According to Araùjo (2010), it is because of tourism's ability to have a direct and indirect impact on any country's economy that the public authorities, both municipal, state and federal, have invested so vigorously in this sector of the economy.

Taking the aforementioned information as a reference, it is necessary for tourism enterprises to include environmental actions and practices at all levels, so that the environmental impacts caused by economic activities on nature's resources can be reduced as much as possible. In other words, sustainable planning can be carried out in tourism enterprises.

2.1.1 Origins of sustainable development or sustainability

It is not new that human activity, with its technological advances, has had a negative impact on the environment and, consequently, on the natural resources available. Resources that until recently were considered inexhaustible.

According to Cavalcanti (2003), after the agricultural (18th century) and industrial (from 1760) revolutions, man became increasingly dependent on the resources that nature made available, and nature became the object of manipulation and transformation by man in order to serve the interests of humanity. As such, it can be said that the ecological crisis is the result of human development, carried out in a disordered manner.

Society is constantly growing in terms of consumption, requiring more and more immeasurable means of production, as well as logistics and waste management. These means are outstripping the planet's finite capacity (DIAS, 2008).

The increase in production capacity has increased the amount of waste generated, especially since the Industrial Revolution, which has led to the emergence of a variety of substances and materials that did not exist in nature (BARBIERI, 2007).

As mentioned earlier, the development of humanity is linked to the degradation of the environment, caused by the unconscious exploitation of natural resources. Based on this information, around the 19th century, society showed its first signs of concern for the planet through demonstrations aimed at creating national parks (COOPER, 2007).

The realization that natural resources are exhaustible and that it would be necessary to take the environment and society into account in the production of goods and services sparked concern about possible solutions that would encompass social and environmental development as well as economic and cultural balance. This was later conceptualized as Sustainable Development. The evidence was becoming increasingly clear that if there was no sustainable environmental awareness, the physical environment and quality of life could suffer great losses or even reach complete destruction (CARDOSO, 2005).

Sustainable Development, from the point of view of organizations, mainly seen as a set of discourses made in the global context, can directly interfere with organizations in a strategic way. Various authors have dealt with this issue, such as Cardoso (2005); Camargo (2002); Cooper (2007); Tachizawa (2008); Deery and Fredline (2005), among others.

According to Camargo (2002), the concept of Sustainable Development was first suggested by studies carried out by the United Nations Organization (UNO) on climate change in the early 1970s. The purpose of this study was to address humanity's concerns in the face of the environmental and social crisis that has engulfed the world since the second half of the last century.

Since then, a series of events have been triggered, bringing together countries from all over the planet, each with proposals and objectives to minimize the impacts on the environment caused by their economic and social growth.

To make things easier to understand, below is Table 01, which refers to the main global events, as well as reflecting the concern of nations and government bodies about the environment and sustainable development.

Chart 1 - Historical milestones in environmental education.

Events	Year	Description	Objective
Launch of the book "Man and Nature"	1864	-	It aims to select and define the studies to be carried out on the relationship between species and their environment.
Yellowstone	1872	The creation of the world's first national park.	Environmental awareness.
Creation of the Club of Rome	1968	They are influential people who come together to debate issues of politics, economics and, above all, the environment and sustainable development.	Promote understanding of the varied but interdependent components - economic, political, natural and social - that make up the global system.
Stockholm Conference	1972	It was the first major meeting organized by the United Nations to focus on environmental issues.	Establishing national environmental legislation to control environmental pollution.
Brundtland Report	1987	The result of a study, the fruit of a 1980 UN convention.	It proposes sustainable development.
ECO 92 or Rio 92	1992	It definitively enshrined the concept of sustainable development.	Reconciling socio-economic development with the conservation of the planet's ecosystems.
Creation of the Commission for Sustainable Development (CSD)	1992	Subsidiary body in environmental matters, subordinate to the Economic and Social Council. UN Social Council (ECOSOC).	First and foremost, to ensure the continuation of the objectives set by the Rio Conference.
World Summit on Sustainable Development or Rio + 10	2002	An event held in South Africa, which brought together several countries from around the world.	Its main objective was to establish an implementation plan that would accelerate and strengthen the application of the principles approved at Rio 92.

Kyoto Protocol	2005	ªThis was an environmental agreement reached during the 3rd Conference of the Parties to the Convention on the Law of the Sea. of Nations United Nations Framework Convention on Climate Change.	Reduce emissions of polluting gases.
Rio + 20	2012	Rio + 20 is the name given to the United Nations Conference, whose premise was to deal with issues related to Sustainable Development.	To renew and reaffirm the participation of country leaders in sustainable development on the planet. It was therefore a second stage of the Earth Summit.

Source: adapted from Dias (2011).

Cooper (2007) points out that responsibility for the sustainability of the planet lies not only with governments and international organizations, but also with industries and their respective consumers.

Following this line of reasoning, Tachizawa (2008) adds that the new economic context is characterized by customers who are increasingly more rigorous when it comes to products and services that have environmental certifications and who give preference to companies that have a good image in the market in terms of maintaining an ecologically correct stance.

The term "Sustainable Development" has been bandied about a lot, but what is this new terminology that has become increasingly present, especially since the 1990s, both in national and international conferences and in environmental awareness plans and in the routine of new managers? Sustainable Development, according to Swarbrooke (1998), consists of development that meets the present needs of society without compromising the ability to meet future needs.

This definition links environmental, economic and socio-cultural objectives, the so-called three pillars of the *triple bottom-line* approach to sustainability (DEERY; FREDLINE, 2005).

Figure 01 illustrates what the author refers to as the three tripods of sustainability.

Figure 1- Sustainability tripod.

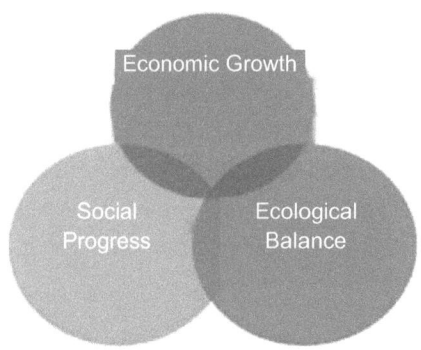

Source: Deery and Fredline, 2004.

For the author, the concept of sustainable development refers to the combination of achieving economic growth, but with social responsibility and care for the environment. The term draws attention to an alternative to theories and signals a warning to traditional models of development, worn out in an endless series of frustrations.

Sustainability, on the other hand, means the possibility of continuously achieving equal or even superior living conditions for a group of people and their successors in a certain ecosystem. In other words, this concept is equivalent to the idea that our life support system can be maintained. In a nutshell, it's about recognizing what is biophysically possible from a long-term perspective (CAVALCANTI, 2003). The author adds that the development experienced by the world over the last two hundred years, especially after the Second World War (1945), is unsustainable.

Sustainability is defined by Wagner (2005) as a possibility for competitive advantage. Despite different approaches and perspectives on sustainable tourism, one of the important components for achieving sustainable tourism development is the participation of various stakeholders (CHEN et al., 2014). Ko (2005) suggested that the various stakeholders should be involved from the initial stage of the sustainability assessment.

From the above, we can say that the development of sustainable tourism requires three basic principles: (1) compromise between conflicting interests and objectives; (2) cooperation between decision-makers (government), the local community, tourism operators and consumers; and finally, (3) the promotion of the public interest in the long term.

Mebratu (1998) presents an extensive categorization of theories of sustainable development. The author identifies three typologies of the theoretical concept of SD, presented as follows: institutional, ideological and academic. The researcher states that in the institutional version, the conceptions of the Institute for Environment and Development (IAD) and the World Business Council for Development (WBCSD) are discussed, addressing and discussing their guiding questions and defining their objectives. The ideological version presents ecotheology (which is more spiritual in nature), ecofeminism (which is linked to the women's movement) and finally ecosocialism (which has a Marxist approach, focused on labor movements). The last version described by the author is the academic one, divided into three parts: economist, ecologist and sociologist.

Table 02 details the three parts of the academic view of sustainable development presented by Mebratu (1998).

Table 2 - Comparative analysis of the academic version of sustainability.

Academic discipline	Epistemological Orientation	Source of the Environmental Crisis	Epicenter of the Solution	Solution Mechanism
Economy Environmental	Economic reductionism	Depreciation of ecological assets	Internalization of externalities	Marketing tools
Ecology Deep	Ecological reductionism	Human domination over nature	Reverence and respect for nature	Biocentric equality
Social Ecology	Holistic - reductionist	Domination of people and nature	Evolution of nature and humanity	Rethinking the social hierarchy

Source: Mebratu (1998).

It is worth pointing out that we are experiencing an innovative moment, sometimes considered exemplary, because it presents a new world view, which is the view of sustainable development. Organizations are realizing that they can stand out from their competitors and are seeking to implement practices that meet sustainable requirements.

2.1.2 Tourism

The term tourism was coined in the 19th century, but tourism has been around since the dawn of mankind. It has been proven that various forms of tourism have been carried out since the earliest civilizations, but it gained strength in the 20th century, more precisely after the Second World War. The tourism sector evolved as a result of aspects related to business productivity and, above all, the purchasing power that people were acquiring and the well-being that resulted from the restoration of world peace (FOURASTIÉ, 1979).

Tourism is now considered to be one of the main activities contributing to global economic growth. It encompasses an immense variety of activities ranging from recreation, leisure and relaxation to the establishment of a partnership between large companies, which is the case with business tourism, or health care.

Currently, the tourism sector, also known as the "industry without chimneys", has been developing and standing out as a relevant and unique economic activity for the whole world, reaching 9% of the world's GDP in 2011 (WTTC, 2011), a factor that makes the tourism industry interesting for the world economy.

There are various definitions of tourism. For the World Tourism Organization (WTO), (CRUZ, 2001):

> Tourism is a form of spatial displacement, i.e. involving the use of some form of transportation and at least one overnight stay at the destination; this displacement can be motivated by the most diverse reasons, such as leisure, business, congress, health, and other reasons, as long as they do not correspond to a form of direct remuneration (CRUZ, 2001, p. 4).

Fuster (2003) understands tourism as a composite of two significant parts: on the one hand, a group of tourists; and on the other, facts and connections derived from their actions, which produce consequences during their travels. These consequences include the use of natural resources and the generation of waste, which, if not managed properly, can have an adverse impact on the natural and built environment.

For McIntosh, Goldner and Ritcie (2003), tourism is nothing more than the sum of the phenomena and/or relationships that arise from the interaction between tourists, companies providing services (tourist agencies), governments and host communities with the function of attracting and accommodating these visitors.

Since the implementation of this activity, tourism has been recognized by many as a sector of the economy that generates development for the locality that benefits from it. From the point of view of some developed countries, tourism is seen as an industry "without chimneys", which provides the much-dreamed-of jobs and income needed to finance other economic activities (FREITAG, 1994).

Walpole and Goodwin (2000) state that:

> Advocates of this idea cite numerous potential benefits for local communities, including increased revenues, increased availability of jobs, and greater market stability than that provided by the export of *commodities*, which is a term used to refer to raw *commodities* (raw materials), or those with a small degree of industrialization, produced in large quantities. These *in natura*, cultivated or mineral extracted products can be stored for a certain period without significant loss of quality. They have a global price and marketability (they are traded on the stock

exchange). Examples: oil, soybeans and gold (WALPOLE; GOODWUIN, 2000, p. 68).

Tourism thus makes a crucial contribution to the income of the regions where it is developed, and can reach rates of up to 70 percent of total GDP in less favored countries (WTO, 2012).

The tourism industry has more than just supporters. For critical researchers, tourism carried out unconsciously can lead to: (1) problems related to the sector's excessive dependence on foreign capital; (2) inequalities in the distribution of the benefits generated by the sector, especially with regard to the income generated; (3) deterioration of the environment in which the activity is carried out and (4) as well as other harm resulting from tourist activity on the host population (PEARCE, 1991; LIU; WALL, 2006; TUNG & AYCAN, 2008).

According to Wheeller (1991), on the negativity of the tourism segment, the author argues that most of the control and issuance of tourists is located in the developed economies, while only the resorts that are created in the destination countries (developing countries), but a large part of the capital raised, returns to foreign entrepreneurs. The author further reinforces this view by stating that international tourism reflects global economic imbalances and the structural dependence of developing nations, i.e. tourism can perpetuate inequalities between developed consumer nations and developing hosts.

Still on the negative impacts caused by tourism, Table 03 presents information showing the main types of tourism according to their typology and their respective damage (PILLMAN, 1992, apud RUSCHMANN, 1998, p. 61).

Table 3 - Types of tourism and environmental impacts.

Types of Tourism	Main Activities	Impacts
Leisure tourism	Walks, rides, rest, recreation, nature watching, accommodation, communication.	Noise, wear and tear on paths and trails, damage to the landscape and vegetation, erosion of beaches and hillsides.
Sports tourism	Skiing, swimming, boating, taking part in competitions.	Air and water pollution, damage to residential areas, attacks on nature due to the construction of sports facilities and gyms, vandalism.
Business tourism	Doing business, expanding companies, congresses, lectures, seminars, fairs,	Noise, air pollution (industries), material damage (wear and tear).

		training/studies.	
Vacation tourism		Beaches, trips by car, train, plane or ship, accommodation, camping, city tours, visits to cultural sites.	Intensification of traffic on highways, railroads and airports, noise, air pollution, effluents, damage to vegetation, wear and tear on the soil due to the construction of terminals, highways and railroads, monotony in the landscape, accidents, mass tourism.
Health tourism		A walk, rest, healing.	Effluents, consumption of nature, interference in local daily life, awareness of society's needs.

Source: Adapted from Pillman (1992, p.6) *apud* Ruschmann (1998, p. 61).

This dichotomy is explained by Lea et al (1988), who points out that, in modern literature, tourism studies have been divided into two schools of thought: "Politico-economic" and "Functional". The "Political Economy" approach is based on the premise that tourism has developed in a very similar way to the historical patterns of colonialism and economic dependency. According to this view, the industry is so governed by political and economic determinants that little attention is paid to other aspects. Analyses in this approach tend to be negative about the effects of tourism, which is seen as just another economic way for developed and wealthy nations to develop at the expense of the less fortunate.

In contrast, the other view presented by Lea et al (1988) is the "Functional" approach. This emphasizes the economic importance of tourism for all participants and ways of improving its efficiency and minimizing its negative effects, without any involvement of politics. This perspective places little emphasis on the history of change in developing societies and the potential contribution of the tourism industry to the localities present.

For Carvalho (2012), unlike the previous perspective, this one offers an optimistic view of the segment, seeing most of the problems as solvable through management and appropriate policies.

These policies can usually be defined as following a strategy of tourism specialization, or following a line of sustainable planning in their implementation, on the economies and social development present in these regions, due to their strong natural capital in tourism resources (SOUSA; FONSECA, 2013).

According to Korossy (2008), it can be seen that for a long time, the emphasis on this activity was almost exclusively on the economic aspects and the contribution that tourism could make to the Gross Domestic Product (GDP). However, tourism is no longer seen solely as a source of income, but rather as a way of discovering other forms of leisure and relaxation linked to socio-cultural and environmental growth, both for those who do it and for those who receive it (receiving communities, i.e. local regions where tourism is carried out or developed).

The aim is to achieve the desired balance between economic and social development and environmental preservation, which can only be achieved through the consistency and sustainability of development in tourist territories. Rosvadoski-da-silva, Gava and Deboça (2014) contribute by stating that this goal can only be achieved specifically through the combination of two variables: (1) an increase in income and forms of local wealth and (2) while ensuring the conservation of natural resources, with the social standards already established.

In this context, various other forms of tourism are being proposed, such as responsible, alternative, ecological and, most recently, sustainable tourism (DIAS, 2008).

These proposals have been adopted with the aim of minimizing the damage caused by tourism and maximizing the benefits generated by it. Generating sustainability in the locality where the activity is carried out.

2.1.3 Tourism and Sustainability

Sustainable tourism can be described as tourism that takes place in any environment, but aims to be responsible in line with sustainable development. Regardless of the type of traveler experience offered. Tour operators doing business in protected areas must comply with the requirements of natural area managers, for example in terms of the areas they can access, as well as the types of activities and impacts they can offer, and must therefore embrace aspects of sustainability in their operational routines (DIANE; JENNIFER, 2011).

Concern about sustainability applied to tourism activities is not a recent issue (ANDRADE; BARBOSA; SOUZA, 2013). As previously mentioned, Santos; Chaves (2014), reinforce this by stating that tourism has been growing rapidly in developing countries, and in Brazil this reality is no different. This significant growth in tourism has been noticed for decades and, as a result, has been given its due importance.

Tourism has a significant impact on the lives of the people who travel and the local inhabitants of the destination visited. Many concerns about the environment have arisen in recent decades, as not all natural resources are finite and renewable (MEDEIROS; MORAES, 2013).

Regarding sustainability in tourism (MALTA; MARIANI, 2013) make the following statement:

> The growth of today's organizations has led to the adoption of management measures based on environmental and social awareness, with the aim of adding competitive and, consequently, financial advantages. In this context, tourism presents itself as an activity that must be essentially geared towards sustainability, since the surroundings of the tourist site must be attractive enough to be visited and, paradoxically, its development contributes to its externalization (MALTA; MARIANI, 2013, p. 123).

Despite the repercussions of the topic and publications extolling the benefits of applying sustainability to tourist destinations, some authors have criticized the ambiguity of this concept (PANATE, 2015).

Reinforcing the above, Mccool and Moisey (2001, p.3) state that "the meanings attributed to sustainable tourism vary widely, with apparently little consensus among authors and government institutions". Cohen (2002) goes further, warning of a subjective problem and raising the question that the vague nature of the concept of sustainability in tourism leaves room for misuse by interested parties, in particular tourism entrepreneurs, since a company that carries out sustainable practices is well regarded in the market by potential clients.

Still on the subject of the misuse of the term sustainable tourism, Cohen (2002) makes an observation about the use of the term "ecotourism", which is used worldwide, but which tourism enterprises often don't even have concrete initiatives to apply to preserving and caring for the environment. The author insinuates that the same is happening with sustainability.

Given this profusion of concepts, it is natural to suggest ways of categorizing them. Some of them are presented below.

Wheeller (1991) states that sustainable tourism opts for the traveler over the conventional tourist, the individual over the group, prefers the employee over the big company, simple and rudimentary accommodation over the big multinational hotel chains, the small over the big, in other words, the essentially good over the apparently exalting.

The author's comment implies that sustainable tourism is a simpler and more rudimentary form of tourism, where those enjoying the activity have greater contact with the locality and the resident population. Where there is concern about the number of visitors and, consequently, the solid waste generated and the possible damage caused.

Following this reasoning, advocates of alternative tourism demand the total replacement of mass tourism with small-scale tourism (LANFANT; GRABURN, 1992).

Based on this premise, it can be said that the initial understanding of sustainable tourism was bifurcated, with authors clearly understanding sustainable tourism as the domain of a certain type of tourism, based on small-scale characteristics.

The attitude of sustainable tourism is in line with the development of an activity that expresses human awareness of its effects at all times. There is no longer any way of affirming the inexistence of the sometimes negative consequences of practices based simply on economic visions, especially with regard to the environment, recognizing the limitations of the natural resources to be exploited (MEDEIROS; MORAES, 2013).

It is important to make it clear that sustainable tourism is not only about caring for local nature, but also for local society and culture. In this sense, Corsi (2004) states that tourism today has great expectations for the continuous improvement of the lives of communities, where the integration of tourists with the population is smooth and efficient, with great additions of knowledge for those who arrive and stay there for a while, as well as for those who live there.

This definition refers to the different categories of tourism, especially those in which tourists not only consume traditional industrialized products, but also gastronomy, handicrafts, artistic shows and create a bond of interaction with the native people of those regions (CASTROGIOVANNI et al., 2001).

Developing and maintaining sustainable planning is essential for balanced tourism development in harmony with the physical, cultural and social resources of the receiving regions, thus preventing tourism from destroying the foundations that make it exist (RUSHMANN, 2008).

Still on the concept of sustainable tourism and using the definition of Cooper et al (2007), which defines sustainable tourism as the development of tourism that meets the current needs of tourists as well as host regions and, at the same time, guarantees opportunities for the future.

The author goes on to say that sustainable tourism aims to ensure that all resources are managed in such a way that economic, social and aesthetic needs can be met, while maintaining local cultural integrity, essential ecological processes, biological diversity and life support systems.

Sustainability in tourism is complex, as it has to guarantee the long-term preservation of the environment, as well as ensuring that those who invest in tourism see a return on their capital and growth in the company's own results. Sustainable tourism, in the long term, has to be ecologically durable, economically viable, but also socially and ethically fair to the local population (DAVID, 2011).

Sustainable tourism is a process that aims to minimize as much as possible the tensions and friction that exist between the complex interactions provided by the tourist *trade*, which is the set of superstructure facilities that make up the tourist product. These include accommodation facilities, bars and restaurants, convention centers and trade fairs, travel and tourism agencies, transport companies, *souvenir* stores and all the peripheral commercial activities linked directly or indirectly to tourism, in other words, reducing the conflicts that may exist between the entities that make up tourism: visitors, the environment as a whole and the local communities that receive tourists. It is therefore a perspective that involves striving for the long-term viability and quality of natural and human resources, i.e. producing tourism, developing the local community, without damaging the environment or the culture of the receiving society (GARROD; FYALL, 1998).

Sustainable tourism therefore proposes to implement a fairer division of coexistence between tourism and the environment, without either partner suffering harmful consequences, seeking a balance between economic issues and environmental conservation (CORSI, 2004).

According to the National Tourism Confederation (CNTur), Brazil is still in its infancy when it comes to sustainable tourism. It is taking its first steps towards developing this type of tourism, and there is still a greater interest in offering ecotourism (which, although it may seem like it, is not aimed at preserving the environment, but only at enjoying it). However, some large national tourist centers, such as Bonito/MS, have already adopted and are developing sustainable tourism practices, after realizing the need to impose rules so that their natural assets are not destroyed as a result of unplanned tourism (CNTur, 2011).

It can therefore be said that sustainability in tourism involves two processes, one of recognition and the other of responsibility. The recognition that the resources used to make tourism products are expensive and vulnerable. And the responsibility when it comes to the intelligent use of these resources falls on all stakeholders, from governments and planners, to the sector that provides services, to tourists and local residents (COOPER, 2007).

Thus, tourism development must be planned taking into account the balance and equity between the dimensions of sustainability, otherwise it can lead to many negative impacts on social and environmental sustainability for the locality that develops it (SANTOS; CHAVES, 2014).

In this way, the development of sustainable tourism involves managing not only natural resources, but also maintaining the human habits and customs of the local society where tourism is taking place, so as to bring pleasure to visitors and, at the same time, benefit the locality while minimizing the negative impacts on the region and the resident population.

According to Diane and Jennifer (2011), in several regions of the world, especially in developed countries, there is a requirement for the development of partnerships between the tourism industry and protected area management. However, the underlying objectives of these partnerships are somewhat different when compared to the actual actions carried out in protected areas, with the operators focusing on biodiversity conservation and tourism *versus* the mission of providing a visitor experience that yields economic profit. Although many of these partnerships have been in operation for considerable periods of time around the world, little is known about their success in terms of approaches to protected area conservation and management, as well as: sustainability, sustainable development and tourism. Although the concept of sustainability is somewhat relative and changeable.

With regard to the thoughts of the aforementioned authors, it makes the reader reflect on how important it is to pay due attention to studies on sustainable development and sustainability, as well as the possibility of new research on the subject in question.

2.1.4 Hospitality

The internal and external pressure on companies to preserve and conserve the environment is forcing changes in the way they think and act, breaking paradigms and creating new ones. Hotels are also in this context (ALVES, 2012).

What is known today about the history of hospitality in the world is that hosting people is a very old practice. The word lodging itself, from the Latin *hospitium*, means hospitality (given or received). And hospitality, also originating from the Latin *hospitalitas*, means the act of offering good treatment to those who are given or receive hospitality.

According to Andrade (2002), the earliest evidence of organized lodging dates back to the beginning of the Olympic Games. This lodging consisted of a large shelter in the shape of a shanty called an *Asylon* or Asylum, which was an inviolable place to allow rest, protection and privacy for outside athletes invited to take part in religious ceremonies and sporting competitions.

Later, with the Industrial Revolution and the expansion of capitalism, lodging came to be treated as a strictly economic activity to be exploited commercially. Hotels with standardized *staff*, made up of managers and receptionists, only appeared at the beginning of the 19th century. According to Chiavenato (2004), *staff* is the result of a combination of linear and functional types of organization, i.e. it is made up of a combination of characteristics of linear and functional types of organization, created in order to combine the advantages of the two organizational styles. The search for a new organizational style to meet the growing efficiency needs of companies led to the creation of this style, which seeks to specialize areas of the organization so that employees' efforts are focused on specific tasks.

According to Petrocchi (2003), the tourism industry, or industry without chimneys, is made up of three basic services: transportation, accommodation and the attraction, with Hospitality and Tourism being an inseparable binomial.

According to Gonçalves (2004), the concept of hospitality began in colonial Brazil, when travelers stayed in the large houses of the mills and farms, in the houses of the cities, in the convents and, above all, in the ranches that existed on the side of the road. The author goes on to say that the Jesuits and other religious orders, moved by the duty of charity, received illustrious personalities and some other not-so-important guests in the convents. It's worth noting, since the subject is about the beginnings of hotels in the country, that in the middle of the 18th century, an exclusive guesthouse building was built at the Sao Bento monastery in Rio de Janeiro.

Also in Rio de Janeiro, in 1908, the Avenida hotel appeared, the largest in the city. In the 1930s, large hotels were set up in the state capitals with casinos as their attraction, but in 1946 casinos were banned. Then, in 1960, with the encouragement of public authorities to boost tourism in the country, the government began to offer measures to encourage the tourist sector.

In the 1970s, with the stimulus of air and road development, Brazil became a target for international hotel chains. At the beginning of the 1980s, projects in the luxury segment were consolidated, as well as the development of economy and mid-range hotels. Despite the international crisis, the 1990s saw an increase in hotel demand in the country (DIAS, 2008).

Regarding the concept of hotels, according to Embratur, in its Normative Resolution 387/98, "Hotel company is: a legal entity that operates or manages a lodging establishment and whose corporate objectives include the exercise of hotel activities".

For Castelli (2003), a hotel can be defined as a building with a preferably urban location, usually with more than one floor, and which offers accommodation and some leisure and business facilities for temporary visitors. In addition to having housing units (HUs) with private bathrooms in at least 60 percent of the housing units, for those already operating.

In modern hospitality, it has become customary to identify the hotel segment as the hotel industry. However, it is not considered correct to identify this segment because the hotel industry is not industrialized, i.e. it doesn't manufacture anything. For example, the hotel industry could be called the accommodation and services industry, since it provides accommodation, food, entertainment and also services.

With the passage of time, and the emergence of environmental awareness on the part of society, tourists have become more demanding, giving preference to products and services that offer environmental prevention measures (DIAS, 2008).

This could lead to the propagation of sustainable tourism itinerary models and the qualification of different destinations, such as beaches, mountains and rural areas, for example. It can also raise awareness of energy rationalization, effluent and waste treatment, among other things (PIRES, 2010).

A hotel is an organization that generates waste of all kinds, which is why it is necessary to implement the concept of environmental management right from the conception phase of the product, the hotel. This concern must start from the project stage, with the planning of an Environmental Management System (EMS), geared to the specific conditions of the location, the preservation of natural resources, the correct disposal of the waste produced and the development of environmental awareness, not only among employees, but also among guests and the community (ALVES, 2012).

This awareness can be achieved in a number of ways, including obtaining "green seals" and becoming environmentally certified through the implementation of environmental management systems.

According to Valle (1995), early concern about the environmental risks that hotel organizations can cause to the environment can generate dividends and shorten the deadlines for the possible acquisition of environmental certifications. Some regulatory and supervisory bodies in certain sectors are already concerned about this issue, as they seek to improve the organization's image by including environmental certification processes in their operational routines. The requirements for these certifications come through laws and regulations.

In view of the above, it can be said that a new generation of tourists is being formed, in which travelers are not only looking for places to go on adventures, rest, do business or feel new sensations. Tourists are increasingly looking for hotel developments that are more concerned with natural resources and that take care to conserve and preserve the environment. This poses a new challenge for managers.

And so, faced with this new global trend, hotel companies must seek out environmental management systems in order to manage the natural resources available in a more effective and rational way due to the possible depletion of these resources, which would lead to a deterioration in the quality of life of the communities where tourism is developed and, consequently, of humanity. The use of ISO 14001 is an excellent strategy that companies can adopt to achieve this goal.

In order to make it easier to read and understand, table 04 presents information as a way of summarizing part of what has been covered in this chapter.

Table 4 - Summary of the literature covered in the chapter.

Theme	Topics	References
Origins Development Sustainable or Sustainability	• Man is hostage to natural resources; • Increase in production capacity X amount of waste generated; • Social and economic balance; development X • Humanity's concern about the crisis environmental; • Main environmental events held in world; • Development that meets society's present needs without compromising the ability to meet future needs.	Giesta (2013) Ivars (2003) Araujo (2010) OMT (2015) Dias (2008) Barbieri (2007) Cooper (2007) Cardoso (2005) Carvalho (2012) Camargo (2002) Tachizawa (2008) Swarbrooke (1998) Panate (2015) Ivanov (2012)
Tourism e sustainability	• Tourism is growing fast in developing countries; • Finite and non-renewable natural resources;	Diane and Jennifer (2011) Andrade; Barbosa; Souza

	• Adoption of management measures based on environmental and social awareness to ensure competitive and financial advantage; • Relative and changeable concept	(2013) Santos and Chaves (2014) Mccool and Moisey (2001) Cohen (2002) Medeiros and Moraes (2013) David (2011) Rosvadoski-da-silva et al (2014) Borges (2011)
Hospitality	• Hospitality (strictly economic activity); • Tourists are becoming more demanding when it comes to enterprises that have environmental qualifications; • The hotel sector requires new management methods • Compliance with certifications and obtaining seals	Andrade (2002) Petrocchi (2003) Gonçalves (2004) DIAS 2008 Castelli (2003) Pires (2010) Alves (2012)

Source: Research data (2015).

Tourism, which has come to be considered one of the most respectable sectors of the economy in the world, responsible for moving significant numbers in the world economy, has been the object of attention in relation to its potential contribution to the development of various communities. Attention is also being paid to sustainability, with the intention of minimizing the environmental, socio-cultural and economic impacts that the activity can cause.

Agents from all areas of the tourism sector are increasingly concerned with achieving and making public a correct performance in relation to sustainability, managing and rethinking the impact of their activities, products or services offered, taking into account their policy and objectives of developing sustainably.

As the hotel industry is a frequently expanding market segment that depends almost exclusively on the attractiveness of a healthy environment, there is a need to add environmental responsibility policies to its social and cultural values.

Hotels that adopt a sustainable approach in their operational procedures seek attitudes and methods that are less damaging to the environment, by re-evaluating their actions and raising awareness among their employees, managers, directors and so on. This is achieved by optimizing the use of material resources, by reusing and recycling waste, by simple ways of rethinking the process and trying to rationalize it. As a result of containing the waste of materials and resources, operating costs are saved, environmental degradation is reduced, and the market opportunities arising from new environmental practices grow. In addition to strengthening the company's image, it can also have a positive impact on employees, increasing the commitment of internal customers as well as the loyalty of external ones, who look for companies that have a positive attitude towards socio-environmental procedures.

2.2 ENVIRONMENTAL MANAGEMENT IN HOTELS

2.2.1 Corporate environmental responsibility

In the second half of the 20th century, with the intensification of global economic growth, environmental problems worsened and began to appear more visibly to large sectors of the population. Particularly in developed countries, which were the first to be affected by the impacts of the Industrial Revolution (DIAS, 2009).

The way in which society's awareness has changed, taking on an ecological profile, has led governments and even companies to adopt a decisive role in increasing the increasingly severe restrictions imposed on old forms of business management.

In this sense, art. 225 of the 1988 Federal Constitution provides the following information.

> Everyone has the right to an ecologically balanced environment, an asset of common use to the people and essential to a healthy quality of life, imposing on the public authorities and the community the duty to defend and preserve it for present and future generations (BRASIL, 2011).

But in practice, this is not the case. In Brazil, environmental management has been characterized by a series of disarticulations between the different bodies involved, a lack of coordination and a shortage of financial and human resources to manage environmental issues (DONAIRE, 2012).

The main factor responsible for the changes and the way in which organizations are held responsible for environmental damage was mainly the natural disasters caused by large companies, which had repercussions in the international media, causing discomfort among the entire population and practically requiring the creation of specific legislation in order to avoid or minimize environmental damage (BARBIERI, 2007).

Due to society's demands for a more consistent and responsible position from organizations, in order to minimize the difference between economic and social results, as well as the ecological concern that has gained significant prominence, and in view of its relevance to society's quality of life, companies have been required to adopt a new position and interact with the environment (TACHIZAWA, 2010).

In short, environmental management has become an important management tool for capturing and creating competitive conditions for organizations, whatever their economic segment (TACHIZAWA, 2010).

Dias (2009) reinforces the last quote, saying that in addition to economic interests, there are internal and external stimuli that can encourage a company to adopt environmental management methods. The internal stimuli are the need to reduce costs, which bring immediate or medium-term financial benefits; an increase in product quality, thus obtaining functionality, reliability, durability and greater ease of maintenance; an improvement in the company's positive image in the eyes of consumers; the need for innovation, in order to differentiate from competitors and maintain an advantage in the market; an increase in social responsibility, acquiring a concern for diversity and the community; raising awareness among internal staff, which directly influences management to adopt corrective or proactive measures in relation to the environment.

The external stimuli are: market demand, which forces companies to improve the way they operate; competition, which leads to better positioning in relation to competitors; public authorities and environmental legislation, which are the strongest controlling factors for companies to adopt environmental management measures; the socio-cultural environment, since the responsibilities that companies assume in terms of a harmonious relationship with nature have grown; environmental certifications, which are a major external stimulus for companies; and suppliers, who influence companies' conduct.

Tachizawa (2010) collaborates by explaining that the trend towards environmental and ecological preservation by organizations must continue permanently. The author states that economic results, together with alternatives that maintain environmental preservation, are increasingly dependent on business decisions that take into account four primary factors:

1 - There should be no conflict between lucrativity and the question environmental;

2 - Environmental movements are growing on a global scale, and they need to be given due attention;

3 - Customers and communities in general are increasingly valuing environmental protection;

4 - Demand, and therefore companies' revenues, are coming under increasing pressure and depend directly on the behavior of consumers who emphasize their preferences for products and services from environmentally friendly organizations.

The first environmental certifications began to emerge from the new scenario facing organizations, in which environmental issues are only increasing.

According to Dias (2003), these certifications refer to levels of care and pre-established standards that companies receive for carrying out their activities in a conscientious manner. These certifications are issued by entities, governmental organizations or not, recognizing that the product offered and the service provided by a given company have met the required environmental standards.

Based on Dias' (2003) assertion, we can conclude that the profile of managers needs to change. Concern about profit is no longer enough. In this case, hotel managers are beginning to feel the need to implement practices in their management systems that meet the requirements of the standards and certifications that are appearing all the time, so that they can maintain a competitive edge in the market, their competitors and customers.

The first environmental labels appeared in the 1940s and were mandatory, as they had to inform consumers of the negative effects of certain products, such as the presence of toxic substances in certain products, for example (KOHLRAUSCH, 2003).

Since then, a series of other labels have sprung up around the world with ecological proposals, such as the Blue Angel, created in Germany in 1978 to label products considered to be environmentally friendly (VALLE, 1995).

The first initiative to establish a Brazilian green label dates back to 1993, when ABNT proposed a joint action to the Environmental Protection Institute (IPA). After the Rio conference, the Financier of Studies and Projects (FINEP) selected ABNT's Environmental Certification Project for products (TACHIZAWA, 2010).

Table 05 shows the main eco-labels that have existed in the world to date.

Table 5 - Main eco-labels in the world.

Stamps	Nationality and Year of Creation	Description
Nordic Swan	Sweden, 1986	It is a green label institutionalized by the Council of Ministers of the Nordic countries, which is administered by the environment agencies of Sweden, Finland, Iceland and Norway.

Blue Angel	Germany, 1987	The Blue Angel (or Blau Engel) is a government seal, an initiative of the German Federal Republic, owned by the Ministry for the Environment, Nature Conservation and Nuclear Safety.
Ecological Choice	Canada, 1988	Canada's Ecological Choice Program (ECP) is an initiative of the Ministry of the Environment. Its coordinating committee includes representatives from public health, consumers, scientists, lawyers, industry and commerce. The technical aspects are the responsibility of the Canadian Standards Association (CSA).
Eco-Mark	Japan, 1989	This green label is administered by the Japan Environment Association and is awarded to products that meet standardized requirements.
NF Environment	France, 1989	The French Green Label is a program designed to certify products that have a reduced negative impact on the environment but offer equivalent performance.
Green Seal	USA, 1990	The US Green Seal is a privately initiated, independent and non-profit organization created in 1990 to set environmental standards for products, product labelling and environmental education in the US.
Environmental Choice	Sweden, 1990	A seal created with the intention of certifying products that are suitable for use and have a lower impact on the environment than comparable products available on the market.
Ecolabel	European Community, 1992	The Ecolabel, the result of a decision by the European Parliament in 1987 and implemented by the Council of the European Union, is a label created in 1992 and reflects a Community environmental labeling scheme. One of its objectives is to adopt a single environmental label in the European Union.
ABNT Quality Seal	Brazil, 1993	The ABNT Environmental Quality Label, from the Brazilian Association of Technical Standards, ISO's representative in Brazil. ABNT participates in the ISO 14000 standards development process as a founding voting member.

Source: adapted from Dias (2009).

These labels are not necessarily intended to distinguish certain products or groups of products as environmentally friendly. Some of these labels have only been created to take advantage of market opportunities, which are becoming increasingly larger and stricter in terms of attention to the environment, reaching more and more significant slices of consumer markets, which prefer products that comply with less impacting and more correct environmental attitudes (DIAS, 2009).

Currently, more than 20 countries form the global "ecolabeling" network, including Brazil, which is being represented through the Environmental Quality seal of the Brazilian Association of Technical Standards - ABNT, ISO's representative in the country.

The importance of implementing environmental certification in the corporate sphere is therefore clear, as there are a number of advantages to be gained, such as: complying with government regulations, meeting the requirements of new consumers, standing out competitively, as well as improving the environmental management system and reducing production operating costs (HARRINGTON; KNIGHT, 2001).

2.2.2 Environmental Management System (EMS)

The environmental issue has become a major issue in many sectors of the economy, since most businesses depend on the environment and its resources to be able to carry out their activities. Unfortunately, the adoption of simple quality programs by organizations is no longer enough to guarantee good results, so companies, especially those in the hotel sector, are increasingly adhering to Environmental Management System (EMS) practices.

An EMS is the part of an organization's management system used to develop and implement its environmental policy and to manage its environmental aspects (ABNT, 2004).

According to Dias (2011), the environmental management system can be defined as a set of organizational responsibilities, actions, procedures, processes and resources that are adopted so that an environmental management system can be implemented in a given company or production unit.

Using Dias' definition of EMS as a hook, and to make it easier to understand, Caon (2008) offers a practical example that explains the purpose of an environmental management system. According to the author, the aim of the EMS is to achieve, control and maintain the level of environmental performance established by the legal standards currently in force and related to sustainable development, such as ISO 14000, for example.

The implementation and certification of EMSs has emerged as a worldwide trend, given the need to act in the face of sustainable development. Although environmental legislation is becoming increasingly stringent, companies are adhering to environmental management systems as a competitive differentiator, supplying products or services through ecologically appropriate processes (GAIA, 2001).

Dias (2011) states that the adoption of an EMS in an enterprise must be accompanied by changes in environmental culture and awareness in all its sectors, leaving behind some habits and customs from the past that do not contribute positively to the new practices adopted. Another important issue that needs to be rethought and worked on at all hierarchical levels.

Gonçalves (2004) describes four types of system that have been implemented in Brazilian hotels. These systems are shown in table 06.

Chart 6 - Types of Systems Adopted in Brazilian Hospitality.

System types	Description
ABIH Environmental System	Hosts of nature: its actions are guided by three principles, with an operational program with 4 phases: 1 - Awareness and adherence; 2 - Training of the entrepreneur and his collaborators; 3 - Development of environmental plans and 4 - Search for environmental certification.
Environmental Cleaner Production System	The search for products that have the least possible impact on the environment in their manufacture.
Autonomous Environmental System	Developed by some hotels or chains to manage water and energy consumption, recycling and/or broader objectives.
Environmental System Based on ISO 14001	This system consists of six stages: 1 - Environmental policy, objective: pollution prevention. 2 - Planning, which aims to define the goals to be achieved. 3 - Implementation and Operation. 4 - Verification, in this phase the EMS's compliance with legal requirements is analyzed. 5 - Management analysis. 6 - Continuous improvement.

Source: adapted from Gonçalves (2004).

It is worth noting that in addition to meeting the requirements of the ISO 14001 standard, organizations that implement the system and seek socio-environmental certification must carry out a process of continuous improvement while complying with the defined environmental policy. This process is known as PDCA (Plan, Do, Check and Act). Translated into Portuguese, it means: Plan, Implement, Check and Act, and its main objective is to monitor, evaluate and make the necessary corrections to maintain the system (PCTS, 2004). This process is simplified in figure 02.

As advantages of the system, an EMS provides order and awareness for organizations to address their environmental concerns, through the allocation of resources, definition of responsibilities and continuous evaluation of practices and processes (ABNT, 1996). In addition, according to Assumpçâo (2009), having an EMS can help the organization to offer confidence to its stakeholders (employees, customers and suppliers).

Figure 2 - Plan, Do, Act and Chec cycle.

Source: PCTS, (2004).

Harrington and Knight (2001) state that there are many advantages to an effective EMS that meets or exceeds the requirements of ISO 14001. Some of these advantages are: management's expectations are clearly communicated to employees; the organization has a much more predictable design; the EMS provides a basis for all organizational improvement activities; the EMS minimizes the amount of errors that occur, as it documents work instructions; it also eliminates the need to "reinvent the wheel" all the time and makes it possible to ensure that the gains from improvement are captured and internalized.

2.2.3 Environmental management in hotel developments

Environmental concerns have been increasing over the years, a concern that also affects hotel businesses, and various actions are being proposed in this regard. An example of this is a partnership between the Brazilian Tourism Institute (EMBRATUR) and the Brazilian Hotel Industry Association (ABIH) for the new hotel classification, in which there is a requirement for these enterprises to be more concerned about the environment in order to obtain the five-star classification, including in their classification processes actions and processes that characterize attitudes that prioritize environmental responsibility

(CAON, 2008).

As an incentive to encourage companies in the hospitality sector to introduce environmental management practices into their activities, in 2006 the government decided to create ABNT, and through the Brazilian Standard (NBR) 15401, it is aimed at lodging establishments and their sustainability management system, aimed at the planning and operation of activities, following the principles established for sustainable tourism. It can be applied to businesses of any size or type. Its legal requirements contain information on the significant environmental, socio-cultural and economic impacts that an enterprise of this type can cause (NBR 15401, 2006).

The aforementioned standard also takes into account legal requirements and contains information on environmental, socio-cultural and economic impacts that are significant as environmental requirements. It also establishes practices for preparing and training for environmental emergencies, such as: fauna and flora; natural areas; landscaping; architecture and the impact of construction on the environment in which it is located, solid waste and effluents; energy efficiency; conservation and management of water use and the selection and use of environmentally aggressive or non-aggressive inputs (NBR 15401, 2006).

Based on this assumption, Gonçalves (2004) adds that in this new business scenario, companies, whether in the hospitality sector or not, are under strong pressure to change. This is the result of the recognition of major issues, such as the environment. These pressures are represented by a range of immediate forces, such as laws, fines, and the profile of new consumers, which force organizations to move towards the environmental era or even go out of business.

To sum up, it is worth pointing out that the new era of companies with an awareness of environmental issues will contribute to the growth of a new environmental culture in these organizations and together with the organizational culture already formed. This could serve as a bridge to a more harmonious relationship with the environment. It is important to raise awareness among those responsible for managing hotel enterprises and their employees, since this is a long-term global process and it is up to everyone to do their part, always looking for improvements and/or refinements in their practices. So that these companies can carry out their processes with the least possible damage to the environment and reach consumers who are increasingly demanding when it comes to sustainable actions.

In order to make it easier to read and understand, table 07 presents information as a way of summarizing part of what has been covered in this chapter.

Table 7 - Summary of the literature covered in the chapter.

Theme	Topics	References
Responsibility Environmental Business	• Pressure exerted by the government to take measures that maximize support for the environment; • "Ecologically" correct companies, labeled with eco-labels. • Environmental awareness;	Donaire (2012) Barbieri (2007 Tachizawa (2010) Kohlrausch (2003) Valle (1995) Harrington and Knight (2001)
Environmental Management System EMS	• Environmental management as a management tool to boost competitiveness; • A set of organizational responsibilities, actions, procedures, processes and resources that are adopted so that an environmental management system can be implemented in a given company or production unit.	Dias (2011) ABNT (2004) Gaia (2001)
Environmental Management in Hotel Developments	• Environmental certifications; • New actions proposed to hotel companies; • New hotel classification.	Caon (2008) NBR 15401 (2006) ISO 14001 (2012) Gonçalves (2004)

Source: research data (2015).

3 METHODOLOGY

According to Martins and Teóphilo (2007), methodology deals with the variables by which reality can be reached and apprehended as a function of science. It is also the object that perfects the methods and criteria used in the study in order to achieve a certain proposed objective.

In order to meet the research objectives, this section presents the methodological procedures used in this scientific study. The choice of these methodological procedures aimed to meet the scope of the research, as well as the object studied. This research used a qualitative approach, with a descriptive purpose.

A case study was used as the research technique, with information collected using a semi-structured interview script, taking into account the methodology used to carry out the study. The type of research, the participants and the environment in which the study would be carried out were taken into account when drawing up the interview script, as well as the instrument used to collect, analyze and process the data. The information collected was interpreted using content analysis techniques.

3.1 TYPE OF RESEARCH

As for the nature of the work, we opted for descriptive research, which, according to Vergara (2007), aims to describe the characteristics of a given population or phenomenon. One of its peculiarities is the use of standardized data collection techniques, such as questionnaires and systematic observation, with the aim of observing, recording and analyzing phenomena or technical systems, without, however, going into the content.

Through this choice, it was possible to identify and obtain data that allowed us to understand managers' perceptions of social responsibility and sustainability practices in the context of the scenario experienced by companies in the hotel sector in the city of Mossoró, Rio Grande do Norte.

The study also used qualitative research, which Creswell (2010, p. 43) defines as "a means of exploring and understanding the meaning that individuals or groups attribute to a social or human problem".

Martins and Teóphilo (2007) argue that the qualitative method is characterized by the description, understanding and interpretation of facts and phenomena, as opposed to quantitative evaluation.

According to Denzin and Lincoln (2006), qualitative scholars emphasize the socially constructed nature of reality, the deep relationship between the researcher and what is being researched, and the obstacles caused by the situations that are influenced by the research. The qualitative study, in the opinion of these authors, makes the researcher the center of the research process and seeks solvable resources for questions that highlight how social experience is created and can thus acquire meaning. Using the socially and historically constructed reality behind the theme of sustainability as a yardstick, as well as the aim of this dissertation, which is to understand the perception of the managers of the hotel chain in Mossoró, Rio Grande do Norte, about the practices of Corporate Environmental Responsibility and sustainability, the qualitative method is the most appropriate method.

In terms of purpose, a case study was used. According to Yin (2005, p.33), "the case study as a research strategy comprises a method that encompasses everything - from planning logic to data collection techniques and specific approaches to data analysis". Complementing this idea, Vergara (2007) states that the case study can be limited to one or a few units, such as a person, family, product, public body, community and one or a group of companies, which is what this research is about. This means that the researcher will be able to get to know the working environments where the phenomenon in question is to be understood.

3.2 CHARACTERIZATION OF THE ENVIRONMENT AND RESEARCH PARTICIPANTS

To meet the purposes of the research, four organizations in the hotel industry in the city of Mossoró, in the state of Rio Grande do Norte, were chosen, taking into account their size (number of residential units), turnover, number of employees and location. In order to protect the confidentiality of the information about each organization surveyed, fictitious names were adopted for each of them: Hotel 1, 2, 3, and 4, respectively.

Hotel 1 is a resort built on a rich mineral province. Its 200,000 square meters are home to 11 thermal pools, gardens and lots of greenery, offering guests a world of leisure in a perfect setting for a healthy stay. Non-guests can also enjoy the thermal waters for an entrance fee. The hotel is located on the urban perimeter of the city of Mossoró. It has 235 employees and 120 rooms. The leisure area includes a restaurant bar, an artificial lake with pedalos and kayaks, a green area, a soccer pitch, sports courts, a gym, squash courts, tennis courts, a games room, a giant water slide, a wet ramp, a wet playground, sand volleyball, abseiling, a walking track, 11 thermal pools and one normal temperature pool. According to its website, the hotel's history is linked to the discovery of oil in the city of Mossoró. This connection is marked by a picturesque fact: when the pools were being filled for the first time. The pumps had been working all night and in the morning the pools were full of oil.

Hotel 2, located 3 km from the center of Mossoró, close to the exit to Fortaleza, has 269 beds, 110 of which are apartments, and is characterized as an International Lodging Standard, according to information on its own *website*. It has 70 employees and implements good service and refinement in the services provided. It offers facilities and rooms for guests with reduced mobility. There is also a *business center* with computers and a 24-hour reception with bilingual receptionists, as well as an outdoor swimming pool, gym and sauna. It does not have any environmental certifications, but it does maintain some environmental preservation measures, including selective collection, electrical automation and the redirection of used water.

Hotel 3, located on one of the main avenues in the city of Mossoró, has 106 apartments for 320 beds, a reception with *cyber café* and library, swimming pool, bar, restaurant with seating for 200, convenience store, cable TV, private parking (covered and closed), laundry, as well as three convention rooms with complete infrastructure and excellent quality services. With a privileged location, it is situated between two of the biggest tourist capitals in the Northeast (Fortaleza and Natal), with an average distance of 260 km between them.

And finally, Hotel 4, also located on one of the main avenues in the city of Mossoró, has 83 apartments and suites in four categories, which differ in terms of physical space, layout and decoration, some with differentials such as a separate social area, kitchen, balcony or private garden, and is the only hotel among those interviewed that is more concerned about the environment, its mission is to guarantee the least possible environmental impact, adopting measures to monitor and continuously improve our processes and activities, aiming for the rational use of natural resources, raising awareness and training our employees, and raising awareness among our clients and partners, according to information on its *website*.

To make it easier to absorb and understand the information mentioned, table 08 below shows the main characteristics of the companies surveyed.

Table 8 - Characteristics of the hotels surveyed.

Hotels	Origin	Number of employees	Number of units	Characterization	Value of Daily
Hotel 1	Initiative	235	120	Complete leisure area;	From

	Public			Restaurant bar; Artificial lake with kayaks and pedal boats; Large green area; Football field; Sports courts; Gym; Squash court; Courts Tennis court; Games room; Water slide; Wet ramp; Sand volleyball court; Jogging track; Big screen with music videos; Little farm; Orchard; Vegetable garden; Horse-drawn carriage rides; Event rooms; Split air conditioning in all apartments; 32" LCD TV in all apartments; Transfer in luxury 16-seater vehicle; Live music shows; Gastronomic festivals.	R$554,00
Hotel 2	Network Hotels	70	110	Outdoor pool; Gym; Sauna; Restaurant; 24-hour bar; Parking; Wi-Fi; Party room; Valet services.	From R$260,00
Hotel 3	Family Business	35	106	Leisure area; Restaurant; Wi-fi; Parking.	From R$119,00
Hotel 4	Family Business	50	83	Organizing tours and excursions; Sending correspondence; Laundry room; Room Service; Parking; Pet accommodation; Car rental; Express check-out; Complimentary Wi-Fi; Messenger.	From R$152,90

Source: research data (2015).

With regard to the profile of the participants in the survey, four managers were chosen (one for each hotel company surveyed), who were also given fictitious names, GH1 to GH4 respectively, to ensure the confidentiality of the information provided. Table 09 shows the following information:

Chart 9 - Profile of research participants (tactical area).

Code	Position	Age	Sex	Education	Length of time with the

					company
GH1	Commercial Manager	42 Years	Male	Specialist	7 Years
GH2	Operational Manager	37 years old	Male	Superior	3 Years
GH3	General Manager	56 Years	Female	Superior	25 years
GH4	Quality Manager	35 years	Female	Specialist	9 Years

Source: research data (2015).

The participants were chosen according to the principle of theoretical saturation, which according to Fontanella, Ricas and Turato (2008, p. 17), "is a suspension of the inclusion of new participants represented in the research when the data obtained becomes redundant". In this way, the inclusion of new interviewees would add little to the work, and was therefore restricted to the number outlined by the author of the study.

3.3 DATA COLLECTION INSTRUMENT

As far as data collection is concerned, according to Rudio (1999), this is the phase of the research that aims to obtain information about the reality being researched. With regard to the data collection instrument, Mattar (1999) states that it is the document through which the questions are presented to the respondents and their answers are recorded.

A data collection instrument is all the possible forms used to relate data to be collected, using any form of administration, such as questionnaires, topics to be followed during an interview, interview scripts and so on (MATTAR, 1999).

After finalizing the theoretical foundation, we moved on to the development of the data collection instrument, which in this research was developed as follows: through guiding questions and indicators for each dimension studied, thus constituting a semi-structured script for the interview with the managers of each establishment surveyed.

With regard to the semi-structured interview, Martins and Teóphilo (2007, p. 86) state that: "The semi-structured interview is conducted using a script, but with the freedom for new questions to be added by the interviewer". Using the position of the aforementioned authors as a parameter, we used targeted questions contained in the protocol for this research, based on the dissertation by Santos (2009), and adapted to the object of this study and the representatives of the research. New questions were added to the semi-structured script to enable them to be interpreted and transformed into content that could be applied in the research.

The interviews were previously scheduled, in person, with each manager, depending on their availability at work. Data was collected in November and December 2015. In the respective companies to which each manager belongs. The interviews lasted an average of one hour each, were recorded electronically and then transcribed reliably into Microsoft Word 2013 so that they could later be analyzed.

3.4 DATA PROCESSING AND ANALYSIS

With regard to qualitative content analysis, Moraes (1999) states that it should not be limited to definition. The author reinforces this by saying that it is very useful for the researcher to try to go further, in other words, to achieve a more in-depth understanding of the content of the messages through inference and interpretation.

Data analysis was based on the studies of Bardin (2004). According to the author, data analysis should be divided into three stages: a) pre-analysis; b) exploration of the material and treatment of the data; and c) inference and interpretation.

The first stage suggested by Bardin is exemplified by Câmara (2013, p.183) as being "a phase of organization. It establishes a work scheme that must be precise, with well-defined procedures, although flexible". In other words, this stage is characterized as a key factor in organizing the material and analyzing how the data will be processed. What should be used, what should be discarded and possibly something that can be redone. In this study, we opted to use interviews, using a semi-structured script, which were reliably transcribed, and their combination constituted the results of the study.

The second phase of Bardin's theory (2004), known as the exploration of the material, is the interpretation of the transcribed interviews (CÂMARA, 2013). This phase is made up of excerpts taken from the interviews, specifically the questions indicated and aligned with the pre-determined objectives.

In the third and final phase of data processing, the moment of inference and interpretation, all the material collected as a result of the interviews employed was analyzed in such a way as to become expressive and adequate information, so that the results could be achieved (BARDIN, 2004).

In order to make it easier to understand the proposed data analysis, Table 10 presents the specific objectives related to the categories of analysis and their respective data collection instruments.

Chart 10 - Categories of Analysis and their specific objectives.

Specific objectives	Categories	Instrument
To analyze and describe the environmental management and sustainability practices implemented by managers in the main hotel enterprises in the municipality of Mossoró, Rio Grande do Norte;	Environmental Management and Sustainability	Questions 1, 3, 4 and 7 in Appendix B.
Understanding the advantages and challenges of implementing environmental practices from the point of view of managers	Sustainability	Questions 4 and 5 in Appendix B.
To learn about environmental education actions and how they are transmitted to the professionals and clients of the organizations studied.	Environmental Education	Questions 8, 9 and 18 in Appendix B.

Source: research data (2015).

The content analysis was carried out over an average period of two months, corresponding to January and February 2016. This phase was based on a detailed interpretation of each category based on the material produced, the interview scripts and the literature specified during the course of the study, in order to provide consistency in meeting the objectives proposed by the research.

4 ANALYSIS AND DISCUSSION OF RESULTS

This section presents the results of the research, explaining environmental management and sustainability practices, the advantages and challenges of their implementation and environmental education actions and their nuances in the hotel organizations investigated.

4.1 ABOUT ENVIRONMENTAL MANAGEMENT AND SUSTAINABILITY PRACTICES IN THE HOTEL ORGANIZATIONS STUDIED

When asked about the involvement of hotel management in the process of implementing and maintaining environmental practices, the following statements were made:

> Those of us responsible for the hotel's management and administration are 100% committed and directly involved in implementing new environmental practices and conserving existing ones (GH1, 2015).
>
> We are directly involved. Any implementation that comes along these lines, our management is directly involved in supporting it, providing support so that the actions are successful (GH2, 2015).
>
> The company implemented a SEBRAE program called "better tourism". Among the stages of implementing the program, there was a part that focused on environmental management, which was about energy efficiency and food safety; so it was all a process that we spent, I think, eleven months implementing. And all the managers were directly involved in this process, including the hotel management (GH3, 2015)
>
> When we try to implement something here, it always comes from us. So: we always have a practice of listening to the opinions of all the staff, but in the environmental area we never get any suggestions, so the only ones we've had, the few we've managed to implement, have come from us (GH4, 2015).

Although they are aware of the importance of policies and practices aimed at environmental management and sustainability, there seem to be few initiatives to implement them in the hotels studied. It is clear from the answers that only GH3 mentioned an explicit proposal for practices in the organization where he works, which he called SEBRAE's "better tourism" program, pointing to better use of energy resources and food safety, a process that lasted about 11 months. The others surveyed showed that they were aware of environmental issues and were willing to listen to suggestions and opinions, but they didn't comment on what policies they had developed in the organizations where they worked.

According to Rushmann (2008), developing and maintaining sustainable planning is extremely important and indispensable for balanced tourism development in harmony with the physical, cultural and social resources of the receiving regions, thus preventing tourism from destroying the foundations that make it exist.

Medeiros and Moraes (2013) show that the attitude of sustainable tourism is in line with the development of an activity that expresses human awareness of its effects at all times. There is no longer any way of affirming the inexistence of the sometimes negative consequences of practices based simply on economic visions, especially with regard to the environment, recognizing the limitations of the natural resources to be exploited. In this way, the hotel companies in the city of Mossoró surveyed need to think beyond the economic issue and start worrying more about implementing programs and practices that involve environmental management and sustainability in the local tourism context, in order to preserve all the activities present in the municipality.

Asked what sets the hotel apart from other hotel companies, the managers revealed that:

> Without a doubt, our differential is our thermal pools, which reach an average temperature of 48 degrees. Our service is also our strong point. We are always praised by our customers for the service they receive from our staff (GH1, 2015).

> Every day we try to improve within the environmental regulatory standards and other standards, so that we can always provide a better service for our customers and for our employees as well, so that this is a place with a good environment, within the regulatory standards, I think one of our differentials is to be seeking quality, to be among the best in our region, city (GH2, 2015).

> I think our differential here is the quality of the service, the main thing, right, because we have other qualities, for example: in food practices, our restaurant is very well regarded, we have good food (GH3, 2015).

> I think that here, the guest enjoys a hotel with lots of green space, lots of open space, but right in the city center. I think that's the difference, you're in a hotel with a farm feel, but in the city center. Close to everything, well located (GH4, 2015).

The reports show that the differentiating factor presented by the hotel companies is not the expansion of green areas, better use of natural resources and environmental preservation, but other factors such as quality and customer service. Only GH4 showed that it is important to invest in green areas, given that most of the hotels in the city of Mossoró are located in urban areas, this would be a great differentiating factor, leading their customers to have a better quality of life and comfort during their stay.

Although the interests of organizations are more focused on other issues, Dias (2009) states that there are internal and external stimuli that can encourage a company to adopt environmental management methods. The internal stimuli are: the need to reduce costs, which bring immediate or medium-term financial benefits, and the sensitization of internal staff, which directly influences managers to adopt corrective or proactive measures in relation to the environment. With regard to external stimuli, the same author highlights market demand; competition; public authorities and environmental legislation, and environmental certifications, which are a major external stimulus for companies; and suppliers, who influence companies' conduct.

These factors reveal the importance that good environmental management can bring to hotel organizations, since a company that carries out sustainable practices is well regarded in the market by potential customers (COHEN, 2002), constituting a competitive advantage and consequently protecting the environment.

4.2 ADVANTAGES AND BARRIERS TO THE IMPLEMENTATION OF ENVIRONMENTAL PRACTICES IN THE VIEW OF MANAGERS

After interpreting the statements made by the managers surveyed, it was possible to see that their knowledge of the motivations, benefits, barriers and competitive edge of the companies surveyed in relation to environmental issues presented uniform opinions, even taking into account the fact that they were different companies.

Regarding the advantages of implementing environmental practices in hotel organizations, the interviewees stated that:

> Tax reductions, such as income tax, for example. After adapting to environmental licensing, we had a reduction in some taxes (GH1, 2015).
>
> Hummm... It's... ummm, cost reduction, the important focus today regardless of the moment we're going through in our economy, quality is... in the service, I think cost reduction comes, it's very linked, to how, for example, we have a control of towels, washing towels, preserving and saving water, which is an indispensable item today, there are other means that we do to... to give quality in this regard (GH2, 2015).
>
> Well, in these practices that we tried in this implementation, we replaced those air conditioners known as "windows" with central air conditioners, and we achieved energy savings. In the apartments we put in those cards that turn off all the electrical equipment when the guest leaves, and we also saw a difference in energy consumption (GH3, 2015).
>
> I think the environment thanks us, right? Like, it's... the reduction in energy costs, we have the green tariff with Cosern that we use every day for two hours the generator at peak times, in addition to making financial savings, we help preserve the environment (GH4, 2015).

Authors such as Barbieri (2007) and Dias (2011) comment that companies gain a competitive edge when they implement an Environmental Management System in their operational routine, and the acquisition of an environmental seal is the best way to differentiate environmentally friendly products and services.

This differs from the opinion of the managers interviewed, who believe that the main advantages of implementing an environmental management system focus on reducing costs.

It can be seen that managers' attention to developing environmental practices is more focused on reducing costs and saving money. Factors such as tax reduction, control of towels, replacement of air conditioning units, use of key cards on doors to turn off electrical equipment, installation of LED light bulbs, presence sensors, among others, contribute to the understanding that the more investment is made in actions aimed at preserving the environment, the greater the savings and reduction in expenses will be, and consequently the efficiency in the use of resources.

As for the challenges of developing environmental practices, here are the comments from the managers:

> Certainly the biggest barriers are raising awareness among employees and customers. Getting people to change their habits and no longer allowing them to take on attitudes that they thought were common throughout their lives (GH1, 2015).

> Raising people's awareness, I think the main thing, when we're going to put an idea into practice, an action aimed at the environment, I think that raising awareness, both of the customer, the user of our service, and of the employees, knowing that separating waste, saving electricity, water, raising people's awareness, in my opinion is an important factor (GH2, 2015).

> There are always barriers, there's always some resistance, you know, there are people who don't believe, but then we keep hitting that key until we show that it's worth it (GH3, 2015).

> Some things like "everything you do to buy more natural things is more expensive". For example, *LED* light bulbs, organic products are always much more expensive than traditional products, so to speak (GH4, 2015).

However, as every implementation has its benefits, difficulties are also encountered, the main one cited by the managers interviewed being the difficulty of making hotel staff aware of the positive practices of environmental preservation, as well as the resistance they have to keep performing their tasks correctly. Another point mentioned was the difficulty in acquiring natural or low-consumption products, due to their cost not being attractive to exchange for more commonly used products.

4.3 ENVIRONMENTAL EDUCATION ACTIONS

With regard to the environmental education actions proposed by the hotel companies investigated, the interviewees revealed that:

> We have notices in the apartments, for example, asking guests to leave a towel in a certain room when they want to reuse it. As for the staff, we also do a task force, which we do every six months, every three months, depending on the period. We do a mutirâo that, this mutirâo is more about raising awareness than cleaning, really. It's more about raising awareness among the staff, we do a task force that cleans the hotel. Picking up cigarette butts, bottle caps, which involves all the employees and makes sure that in the future they don't throw a cigarette cap, any garbage in the garden, for example. (GH1, 2015).

> As far as guests are concerned, we make signs in the room where they stay, which is the apartment, so that they make conscious use of towels, inside the apartment, we have devices that turn off the power after a minute when the guest leaves the room. As for the internal rules of the house, for example: we change a resident guest's clothes every three days, we don't change them every day like in other places, because then I'm saving two sectors, right? For example, water and energy, combining the environmental and the financial (GH2, 2015).

> Yes, we have a sign in the bathroom raising awareness about reusing towels, so that they don't have to go to the laundry all day (GH3, 2015).

> For the last three years, we've been bringing in these people from the NGO to give talks to our employees, because it raises awareness so that they can do it at home too, right? Doing selective collection and things like that (GH4, 2015).

The interviewees showed that there were some educational actions offered to guests and clients of the hotels surveyed. For customers, GH1, GH2 and GH3 stated that there was a proposal to reuse their towels in order to improve water savings from washing them, as well as reducing the use of chemical products. In addition, the use of cards that turn off the power in their rooms when guests leave has also been widely used in these organizations.

With regard to employees, GH1 revealed that his organization works with awareness and cleanliness groups, influencing its employees to gradually stop throwing any kind of waste on the hotel floor. GH2 said that employees are encouraged to wash guests' towels in order to save water. GH3 didn't comment on the matter and GH4 announced that measures are taken, such as talks given by NGOs, so that professionals, based on initiatives taken in their own homes, can also make good use of resources in the workplace.

Engagement between all participants, from management to operational staff, is essential if there is to be a good impact on the company's image. To this end, it is important to include environmental education (EE) practices, as it is one of the key elements in the processes of raising awareness and mobilizing people, so that they can develop the best actions for sustainability that are planned by the organizational leadership (GIESTA, 2013).

4.3.1 Training for employees

Regarding the training offered to employees on environmental practices, the interviewees stated that:

> Yes, we always try to give lectures throughout the year, we have at least one lecture a month, on a subject we choose at the beginning of the year, and we always have the environmental issue as... the theme, right, of the lectures (GH1, 2015).
>
> Boy, we... we're specifically working on this area still, which is a new area, but we're always talking to them, not specific training yet (GH2, 2015).
>
> They do, so we usually give lectures, then we call everyone together, then we give them guidance, as well as warnings about saving water and energy (GH3, 2015).
>
> Yes, we have to do it again, in fact (GH4, 2015).

On this subject, it can be seen that the hotels investigated lack formal training for their employees in relation to environmental issues. 50% of those interviewed said there was no training of any kind; the other 50% said there were monthly (GH1) and sporadic (GH3) lectures for professionals on the subject.

According to Oliveira and Pinheiro (2010), training plays a fundamental role in the environmental management of organizations, as it helps to increase the interest and attention of professionals to the importance of the practices proposed by companies, leading to an improvement in their skills and knowledge about the aspects that directly and/or indirectly affect the environmental performance of the organization, such as the efficient use of water, energy, fuels, the ideal treatment of solid waste, among others; as well as developing leaders who can help to achieve the desired results.

This lack of formal training can lead to a number of problems for the organizations in question, because the lack of knowledge makes it difficult for employees to apply the principles of good environmental management that the organization wants to achieve. It is therefore necessary for managers to be aware of the need to invest more in this type of training, so that their objectives can be achieved effectively.

4.3.2 Methods for reducing water consumption

When asked about methods of reducing water consumption, the answers showed that:

> Yes, we have, we have, the apartments, they have the information so that the customer can avoid unnecessary use. And today, at the end of last year, we also installed a control system in our well, to find out how much water we're using, how much has been saved, so that we can have an idea of this control from now on (GH1, 2015).

> Guests are advised to use towels conscientiously and resident guests' clothes are only washed every three days. All our employees today, I can even say it like this, we're going through a... a water crisis in our region, right, and the hotel today is second to none, from the smallest to the largest company that uses water, we've had, we've been through several difficult times this year, with the lack of rain our well has reduced its flow, CAERN, CAERN itself had a serious problem a couple of months ago and we're really at the mercy of the situation in our region, ask me the question again. In addition to the ones I've mentioned, there are some internal actions, such as reducing the watering of the plants, we reduce the hours of our laundry, which is a sector that consumes a lot of water, so we reduce the hours of our laundry, which is to save more water, eee... these others that I've already told you about (GH2, 2015).

> Only verbally about saving money, but there's no system for that (GH3, 2015).

> Yes, in the guest bathrooms there are taps, the ones with automatic shut-offs, only in the guest bathrooms because I can't put them in the apartments because there's no way. It's bad for the guest (GH4, 2015).

The reports show that the methods of reducing water use control in the hotels studied include: a) informing customers about how to avoid wasting water, as well as checking the hotel's wells (GH1); b) the conscious use of towels by customers, so that water is not wasted every day washing them; reducing the time spent using the laundry facilities (GH2); and c) installing automatic shut-off taps in the guest bathrooms (GH4).

The water supplied to the hotels in general is supplied by the RN state water company, while hotel 1 mentions that it has an artesian well to supplement its supply and hotel 3 buys a well from a farm. In addition, they mention that they do not monitor the standard of potability of the water, as they rely on the water received from the supply company, but hotel 3, because it has standing water around the establishment, has samples taken to laboratories for testing.

Knowing the importance and necessity of water in the organizational environment, the rational management of water use is not only a governmental or public issue, but also a concern for companies (BICHUETI et al., 2013). For Lambooy (2011), a number of factors motivate companies to rationalize water use. The first is the organization's self-interest in reducing costs, which benefits itself; the second is its image in society, which has become more attentive and sensitive to environmental issues.

4.3.3 Methods for reducing energy consumption

Regarding the methods for reducing energy consumption, the results showed that:

> Yes, we participate in the green seal with COSERN and we have two generators in the hotel that are activated at peak times, reducing electricity consumption (GH1, 2015).
>
> Yes, we've already made significant savings on our electricity bill. We made an investment in changing all the light bulbs in the hotel, bulbs that consumed a lot of energy, and we changed them to LED bulbs. And signs to raise awareness among customers who are staying with us and awareness among employees too (GH2, 2015).
>
> Only verbally about saving money, but there's no system for that (GH3, 2015).
>
> Yes, as I've already said about the showers, and in the service areas we have those one-touch lamps, which you touch and it just goes out. And in the social areas we have those lamps with a motion sensor (GH4, 2015).

It can be seen that few actions are taken by the organizations surveyed with regard to electricity consumption. The measures mentioned include using generators at peak times and replacing conventional light bulbs with LEDs.

Replacing equipment that consumes more electricity has been one of the main strategies adopted by today's organizations (BORGES, 2014). Busse (2010) shows that saving energy can generate various benefits for the organizations that do it, avoiding waste and protecting the environment, reducing the risks of environmental impacts such as deforestation, nuclear radiation, rising ocean levels and the greenhouse effect, among others.

However, much still needs to be done to increase the efficiency of energy use by the hotels investigated. Measures that involve, for example, making good use of elevators, making people prefer the stairs on lower floors; avoiding refrigerator and freezer doors being left open unnecessarily; using the maximum capacity of washing machines and dryers in order to avoid double or triple use of these products, among other measures, would help in the process of reducing energy consumption and contribute to better environmental management.

4.3.4 Sewage treatment and solid waste separation

According to the interviewees, the sewage treatment carried out by the hotels studied is as follows:

> Yes, we have one... we were also instructed by one of the city's environmental managers to build our own sewage treatment system, so we built one, because until then we didn't have a basic sanitation treatment system from the city hall, now we do, it's here in the front, we're even going to opt for it, but before we didn't have one. So we built a drain that was approved by the public body responsible (GH1, 2015).
>
> Public sewage treatment. We have no contact with the sewage, it is directly channeled into the public network. Sanitized (GH2, 2015).
>
> The town hall already provides basic sanitation (GH3, 2015).
>
> I don't know (GH4, 2015).

In this respect, all four hotels have shown themselves to be environmentally suitable in terms of sewage treatment. GH1 reported that, even before the city had a basic sanitation system, his organization already had its own treatment system. Today, basic sanitation is the main form of sewage treatment in the organizations studied. As for the separation of solid waste:

> No, not today, the hotel's regular garbage, the town hall collects it, the garbage that comes from construction work, tree pruning, we have a company that collects this garbage (GH1, 2015).
>
> We do. The municipality has selective collection, which comes once a week. We separate it, we try to guide our employees to separate the waste according to the instructions, and the city council, as part of the selective collection, comes once a week to pick up paper, plastic and other items (GH2, 2015).
>
> We have a space here, after the parking lot, which has separate garbage cans so we can separate all the waste according to the standard (GH3, 2015).
>
> As already mentioned, through an NGO that passes through, it collects the waste so that it can be reused (GH4, 2015).

It can be seen that managers are concerned about the separation of solid waste in each of the hotels investigated. They reported that they have spaces for storing, separating and collecting materials from non-governmental organizations, companies and the city council for processing into products such as brooms and paper recycling, although they do not consider environmental factors when outsourcing services.

The importance of separating waste and giving it an appropriate destination is that it brings a series of benefits, such as the production of energy, in order to recover the economic value of these materials; the generation of employment and income; the reduction in the amount of natural resources used for the service provided by hotel companies; and the need to occupy large areas to deal with different types of waste (SEBRAE, 2012).

The authors Trung and Kumar (2005) found that an overnight stay in luxury hotels anywhere in the world could result in between 2.5 and 7.2 kg of solid waste per guest. This causes concern for hotel organizations in Mossoró, as throughout the year, especially in June, July and January, the number of guests tends to increase, generating tons of solid waste.

According to SEBRAE (2012, p.9), "whatever the waste, there will always be a more appropriate destination for it than simply disposing of it. From reuse to energy generation, everything has value and can even become a source of income and a vector for new businesses". Therefore, this statement reinforces how important it is to have a system for processing and/or reusing solid waste in hotel developments.

4.3.5 Environmental legislation

In terms of environmental legislation, the managers interviewed both said that they were aware of the legislation, mainly due to the ease of today's media, but no manager was able to name at least one environmental law in force, and no hotel surveyed has any environmental certification or green seal. Only hotel 3 has an IDEMA license.

This is in line with the studies carried out by Ferrari (2006, p. 48), when the author states that "this data reveals a major gap in the lodging sector, i.e. a lack of knowledge of environmental legislation. It is important that environmental actions are based on legislation".

From this perspective, managers' knowledge of legislation is essential for the effective implementation of an EMS, i.e. without knowledge on the part of managers, the likelihood of these actions being implemented is minimal. According to Ferrari (2006), the efficiency of a hotel can be improved by optimizing the use of its inputs, reducing waste and residues, as well as complying with environmental legislation.

Finally, we asked about the advantages of a hotel company attracting guests by including sustainable measures in its operating procedures and, as a competitive advantage, we observed in the words of the managers below that two of them, hotel 1 and 3, believe that maintaining sustainable practices will make a difference in the market they are in.

> It's, as I've already said, cost reduction, which is extremely important for the

> company today, the company's image, it becomes more attractive in the market because people are becoming more aware of it, so if the customer thinks: "oh if that company takes care of its waste, then I'll go to that hotel". It's a differentiator from the competition (G1, 2015).
>
> Since everyone has this culture or this environmental education, the advantage it has I think is in its own favor, right, the advantage is this, that it will look good. I don't think it's a competitive advantage (G2, 2015).
>
> I think that guests in general are starting to worry about this issue. I think that if you have to make a choice and there's no difference in the price of your competitor's rate, I imagine that guests will decide to go to the establishment that adopts measures to protect the environment.

This is in line with Borges et al (2010), who states that, in order to achieve greater competitiveness among companies in the same market segment, organizations are using mass media to make their various sustainable development practices known to the public with which they relate. In this way, their sustainability reports and websites provide information to customers, shareholders, investors, speculators, suppliers, employees and the community in general not only about compliance with current legislation on environmental issues, but also about what is practiced beyond what is required by law, such as sustainability practices applied in their operational routines, for example. This responsible attitude ensures a good image for organizations, preserves the environment and, above all, gives them a competitive edge over their competitors.

4.3.6. Skills, knowledge and attitudes needed to care for the environment

The managers interviewed believe that the skills, knowledge and attitudes needed to take care of the environment are based on, among others: low-waste production practices, the use of returnable bags, selective collection, saving water and energy and, above all, the awareness of human beings to take care of the environment, so that they have the necessary conditions to live healthily in harmony with a stable environment.

It can be seen from the managers' thoughts that none of them emphasized environmental certifications or even delved into sustainability or environmental responsibility, citing everyday practices or techniques as an example. They also focused on simpler and more rudimentary ways of preserving the environment, such as saving water and energy, for example.

In view of the results and discussions obtained during this chapter, table 11 is presented as a way of making it easier to understand everything that has been explained in this section.

Chart 11 - Summary of the results obtained in the survey.

Categories	Analysis based on managers' perceptions
Environmental management and sustainability practices applied in the hotel organizations studied	Although managers are aware of the importance of environmental and sustainability policies and practices, according to what was gathered through the interview scripts, there are few initiatives regarding their implementation in the participating hotels.
Advantages and barriers to implementing environmental practices	In terms of advantages, it can be seen that managers' attention to developing environmental practices is more focused on reducing costs and saving money. Factors such as tax reductions, reductions in water and electricity consumption. Regarding the existing obstacles, it was possible to see that the main one is the difficulty in making hotel guests and employees aware of the positive practices of environmental preservation, as well as the costly process of implementing actions aimed at the environment.
Environmental education actions	It was possible to conclude that among the main actions to raise environmental awareness according to the managers are: notices distributed in the hotel premises, with information that proposes saving water and energy, as well as the proposal to reuse towels in order to save water when washing them, and to reduce the use of chemical products. In addition, the use of cards that turn off the power in rooms when guests leave has also been widely used in these organizations.
Training for employees	In relation to possible training for employees, it was noted that the hotel organizations investigated lack formal training for their employees in relation to environmental issues. Half of the managers surveyed said that there was no training of any kind, while the other half said that there were monthly and/or sporadic lectures for professionals on the subject.
Sewage treatment and separating solid waste	Sewage treatment at the four hotels is carried out exclusively by the sewage system provided by the town hall. With regard to solid waste management, the managers interviewed said that they maintain spaces for storage, separation and collection by non-governmental organizations for the processing of products such as brooms and paper recycling, although they do not consider environmental factors when outsourcing services.
Environmental legislation	When it came to environmental legislation, all the managers interviewed said that they were aware of the legislation, mainly due to the ease of today's media, but they were unable to provide any information on any type of environmental law. The hotels do not have any kind of environmental seal or certification.
Skills, knowledge and attitudes needed to care	The managers interviewed believe that the skills, knowledge and attitudes needed to take care of the environment are based on, among others: low-waste

for the environment	production practices, the use of returnable bags, selective collection, saving water and energy and, above all, human awareness of caring for the environment, so as not to have polluted air, polluted water, lack of food, among others.

Source: Research data (2015).

The growing importance given to the socio-environmental policies of companies by consumers is notorious and requires the hospitality sector to adopt new attitudes towards this niche market, which is increasingly gaining ground. New marketing strategies must link the image of hotel companies' socio-environmental initiatives, since none of the companies studied mentioned their sustainable practices in their advertising.

In addition to hotels, with their guest evaluation questionnaires, it is also up to internet sites that evaluate lodging facilities and tourist destinations to add the socio-environmental question to their evaluations, as a way of making guests and tourists aware of the importance of sustainable tourism practices.

Among the strategies used by organizations, including service providers in the hospitality industry, in order to stand up to the competition and remain competitive in the market, is the constant search for innovation, which can be achieved through environmental practices and socio-educational measures aimed at maintaining and preserving the environment.

Innovation is the technical, economic and feasible way of solving a given problem (ZAWISLAK; GAMARRA, 2015).

Various innovative methods, including planning processes, among them planning that prioritizes environmental issues, have been recommended as tools for improving decision-making, and are useful for dealing with the change and uncertainties of an increasingly demanding market. These processes aim to improve decision-making by helping managers to broaden their horizons, recognize, consider and reflect on the uncertainties they are likely to face.

5 FINAL CONSIDERATIONS

This research was essentially a qualitative and descriptive study, as is the case with case studies, and therefore generalizable at a theoretical level rather than at a population level. As such, this study has led to interpretations of the environmental management and sustainability practices applied in hotel enterprises in the city of Mossoró in the state of Rio Grande do Norte. These analyses could be transformed into hypotheses to be refuted in a future study aimed at a population-based sample.

As explained throughout this study, the aim of this research was to identify the perception of managers of hotel organizations in Mossoró, Rio Grande do Norte, about Corporate Environmental Responsibility and sustainability practices. It also had the following specific objectives: to analyze and describe the environmental management and sustainability practices implemented by managers in the main hotel enterprises in the municipality of Mossoró, Rio Grande do Norte; to ascertain the advantages and challenges of implementing environmental practices from the managers' point of view; to find out about environmental education actions and how these are transmitted to the professionals and clients of the organizations studied.

As for the first specific objective, there seemed to be few initiatives regarding the implementation of environmental practices in the hotels studied. With the exception of GH3, who spoke of an explicit proposal for practices in the organization where he works, which he referred to as SEBRAE's "better tourism" program, the other respondents spoke only implicitly about the measures taken, demonstrating that they are attentive to environmental issues and were willing to listen to suggestions and opinions, but did not comment on what policies they have developed in the organizations where they work.

With regard to the advantages and challenges of implementing environmental practices, it was found that managers are more focused on reducing costs and saving money. Factors such as tax reduction, control of towels, replacement of air conditioning units, use of key cards on doors to turn off electrical equipment, installation of LED lamps, presence sensors, among others, were pointed out as benefits of investing in such policies.

In terms of barriers, the interviewees pointed to the difficulty of making hotel staff aware of positive environmental preservation practices, as well as their resistance to carrying out their tasks correctly. Another point mentioned was the difficulty in acquiring natural or low-consumption products, due to their cost not being attractive to exchange for more commonly used products.

In relation to the third specific objective, several environmental education actions were highlighted by the respondents during data collection, among which we can highlight the proposal that customers reuse their towels in order to increase water savings from washing them. In addition, the use of cards that turn off the power in rooms when guests leave has also been widely used in these organizations. Actions are also aimed at employees, such as awareness and cleanliness groups, influencing professionals to gradually stop throwing any kind of waste on the hotel floor; washing guests' towels every three days in order to save water; and lectures given by NGOs so that professionals, based on initiatives taken in their own homes, can also make good use of resources in the workplace.

In addition to these points, it was also found that the hotels adopt measures to reduce water and energy consumption, as well as sewage treatment and solid waste separation. The managers stated that they were more concerned with saving water and energy, as these are the most perceptible items for the hotel in terms of cost reduction. To this end, they maintain a number of sustainable practices, such as the implementation of key cards in the apartments and reminders to guests to save energy, as well as changing incandescent light bulbs to *LED bulbs*.

With regard to reducing water consumption, in addition to visual warnings, they take turns changing the bed and bath linen of resident guests, as a way of reducing the energy used as well as reducing water consumption during washing. In addition, they claim to dispose of solid waste through selective collection and partnerships with non-governmental organizations that collect waste for recycling.

With regard to the evolution of legal norms aimed at sustainability in the hotel industry, it is clear that there have been regulatory advances in favor of socio-environmental benefits, as exemplified by NBR ISO 14001, especially given the tendency of Brazilian legislators to mirror the examples of pioneering countries with regard to policies aimed at sustainable development or sustainability.

The conclusion is that the hotel organizations investigated still have a long way to go in terms of environmental practices aimed at sustainability. Few explicit policies have been adopted so far, showing that environmental management initiatives by these companies are still lacking. It was also noted that although there are educational actions in favor of the environment, these have proved to be timid, especially with regard to saving water, a resource that is becoming increasingly scarce around the world.

Comparing the data collected and analyzing it, it can be seen that the hotels' main and only objective in their environmental practices is to reduce operating costs, and that they do not see environmental education actions as a means of showing themselves to be competitive in the hotel market.

It was also possible to observe that culture, which is included in the social dimension of sustainability, was neglected by one hundred percent of the hotel companies surveyed. The managers of the companies surveyed have minimal knowledge of environmental issues, and the process of implementing environmental practices in the city's hotel companies is still in its infancy. They need to create environmental management systems (EMS), or even innovate the rudimentary practices they have implemented.

The hotels interviewed are companies that are already established in the market and have a certain stability, which means that they don't feel the need to reinvent themselves sustainably in order to obtain new customers, or even to guarantee the permanence of the older ones.

Finally, future research on the subject is suggested, involving a larger number of hotel organizations, as well as comparing hotels in the city of Mossoró with larger ones, in order to understand the differences in environmental practices between them. Quantitative studies are also recommended to measure variables that meet the challenges of effective environmental management in companies.

It is worth mentioning that there is a lack of qualified technical staff to monitor compliance with the country's rules on environmental responsibility, as well as stepping up monitoring of the legal requirements placed on lodging establishments. These are extremely important factors that require more objectivity and investment on the part of the public authorities in order to meet the demand from existing businesses and the new ones that are springing up all the time.

There is also a need for further research into the importance of organizations investing in environmental issues in the tourism sector, given that it is one of the main drivers of Brazil's economic growth, attracting thousands of tourists from all over the world every year.

6 REFERENCES

ABREU, Dora. "**The illustrious green guests**". Salvador, Bahia: Casa da quality, 2001.

ABNT. Brazilian Association of Technical Standards. **NBR ISSO 14004**. Environmental management systems: general guidelines on principles, systems and support techniques. Rio de Janeiro, 1996.

. **NBR ISSO 14001**. Environmental management systems: requirements with guidelines for use. 2. ed. Rio de Janeiro, 2004.

ALVES, Antônio Româo. Environmental management system as a business strategy in the hotel industry. **Revista Produçao**. ISSN 1676 - 1901 / Vol. VIII / Num. III / Santa Catarina, 2012.

ANDRADE, José Vicente de. **Turismo**: fundamentos e dimensôes. 8. ed. Sâo Paulo: Atica, 2002

ANDRADE, M. B. de; BARBOSA, M. de L. de A.; SOUZA, A. de S. Socio-environmental sustainability in the identity of the Fernando de Noronha archipelago and its influence as a factor in tourism promotion. **Revista de investigación en Turismo y Desarrollo local**. Vol 6, N° 14, p. 1-18, Jun, 2013.

ARAÙJO, L. M.; BRAMWELL, B. Stakeholder assessment and collaborative tourism planning: the case of Brazil's costa dourada project. **Journal of Sustainable Tourism**, v.7, 1999.

ARAÙJO, Josemery Alves. **Public policies and the socio-spatial transformations related to tourism in the municipality of Caicó: an analysis of the period 2000 to 2010**. 2010. 147 f. Dissertation (Master's Degree in Tourism and Regional Development and Tourism Management) - Federal University of Rio Grande do Norte, Natal, 2010.

BARBIERI, José Carlos. **Corporate environmental management**: concepts, models and tools. São Paulo: 2007.

BARDIN, L. **Content Analysis**. 3. ed. rev. and current. Lisbon: Ed. 70, 2004.

BICHUETI, R. S. et al. Strategic management of water use in mineral sector industries. In: VI Meeting of Strategy Studies, Bento Gonçalves, 2013. **Anais...**, EEs, 2013.

BORGES, Fernando Hagihara. The environment and the organization: a case study based on the positioning of a company in the face of a new environmental perspective. Dissertation (Master's Degree - Postgraduate Program and Area of Concentration). Supervisor Prof. Dr. Wilson Kendy Tachibana. São Paulo, 2011.

BORGESA, Ana Paula; ROSAB, Fabricia Silva; ENSSLIN, Sandra Rolim. Voluntary disclosure of environmental practices: a study of large Brazilian pulp and paper companies. **Revista Prodçâo**. Santa Catarina, 2010.

BRAZIL. **Constitution of the Federative Republic of Brazil**: promulgated on October 5, 1988. 1988. Available at: <www.planalto.gov.br/ccivil_03/constituicao/constituicao.htm>. Accessed on: July 9, 2015.

BUSSE, B. N. **Academic text on energy efficiency**: a quantitative sample of the last 40 years of research. Available at: < http://www.ipog.edu.br/uploads/arquivos/643a591f20914f664adfe660f87903e5.pdf>. Accessed on 09 Jan. 2016.

CAMARGO, A. Governance for the 21st century. In: TRIGUEIRO, A. **Meio ambiente no século 21**: 21 especialistas falam da questão ambiental nas suas áreas de conhecimento. Rio de Janeiro: Sextante, 2002.

CAON, Mauro Correia. **Environmental Management in Hotels**. 2. ed. São Paulo: Atlas, 2008.

CARDOSO, Roberta de Carvalho. **Social dimensions of sustainable tourism: A study on the contribution of beach resorts to the development of local communities**. Sâo Paulo: FGV, 2005. 264f. Thesis (Doctorate). Escola de Administraçâo de Empresas de Sâo Paulo, Fundaçâo Getûlio Vargas, Sâo Paulo, 2005.

CARVALHO, P. Determining factors of international business tourism: A literature review. **XXII Jornadas Luso-Espanolas de Gestión Cientifica**, Vila Real. 2012.

CASTELLI, Geraldo. **Hotel management**. 9. ed. Caxias do Sul: Educs, 2003.

CASTROGIOVANNI, Alencar C. et al. **Turismo urbano**. 2. ed. Sao Paulo: Contexto, 2001.

CAVALCANTI, C. **Sustentabilidade da economia: paradigmas alternativos de realizaçao econômica**. Sao Paulo: Cortez, 2003.

CAVALCANTI, M. **Gestao social, estratégias e parcerias**: redescobrindo a essência da administraçao brasileira de comunidades para o terceiro setor. Sao Paulo: Saraiva, 2006.

CHEN, Yin; HUANG, Zhuowei; CAI, Liping A. "Image of China tourism and sustainability issues in Western media: an investigation of National Geographic", **International Journal of Contemporary Hospitality Management**, Vol. 26 Iss: 6, pp. 855 - 878, 2014.

CNTUR, National Tourism Confederation. **Sustainable tourism**. Available at: <http://www.cntur.com.br/turismo_sustentavel.html>. Accessed on: June 23, 2015.

CoHEN, Erik. **Rethinking the sociology of tourism**. Annals of Tourism Research, v.6, n.1, 1979.

COOPER, Cyrus. et al. **Tourism principles and practices.** 3. ed. Porto Alegre: Brookman, 2007.

CORSI, E. **Historical and cultural heritage**: a new perspective for urban and rural areas through sustainable tourism. Uberlândia: Caminhos da Geografia v. 5, n.11, 2004.

COUTINHO, Leandro. Especial cidades medias, aonde o futuro já chegou. **Veja**, Sao Paulo: n. 2180, 01 Sep. 2010.

CRESWELL, John W. **Research Project**. Porto Alegre: Artmed, 2010.

CRUZ, Rita de Cassia Ariza. **Introduction to the geography of tourism**. Sao Paulo: Roca, 2001.

DALFOVO, Michael Samir; LANA, Rogério Adilson; SILVEIRA, Amélia. Quantitative and qualitative methods: a theoretical review. **Revista Interdisciplinar Cientifica Aplicada**, Blumenau, v.2, n.4, p.01- 13, Sem II. 2008.

DAVID, L. Tourism ecology: towards the responsible, sustainable tourism future. **Worldwide Hospitality and Tourism Themes**, Vol. 3 Iss: 3, pp.210 - 216, 2011.

DEERY, Gold Coast; FREDLINE, Jago L. **CRC for Sustainable Tourism**, L. A. framework for the development of social and socioeconomic indicators for sustainable tourism in communities. 2005.

DENZIN, N. K.; LINCOLN, Y. S. The discipline and practice of qualitative research. In:

DENZIN, N. K.; LINCOLN, Y. S. **Planning qualitative research:** Theories and approaches. Porto Alegre: Artmed, 2006.

DIAS, Reinaldo. **Environmental Marketing:** Ethics, Social Responsibility and Competitiveness in Business. Sao Paulo: Atlas, 2008.

. **Environmental management**: social responsibility and sustainability. Sao Paulo: Atlas, 2009.

. **Gestao ambiental**: responsabilidade social e sustentabilidade. 2. ed. Sao Paulo:

Atlas, 2011.

DIANE, Lee; JENNIFER, Laing. **Environmental Management**, Vol. 48 Issue 4, p 734 - 749. 16 p. Oct, 2011.

DIEHL, Astor Antonio. **Research in applied social sciences**: methods and techniques. Sao Paulo: Prentice Hall, 2004.

DIEHL, Astor Antonio. **Research in applied social sciences**: methods and techniques. Sao Paulo: Prentice Hall, 2004.

DONAIRE, Denis Jùnior. **Gestao ambiental na empresa**. 2. ed. Sao Paulo, Atlas, 2012.

EMBRATUR, Brazilian Tourism Institute. **Hotels**. Available at: <http://www.embratur.gov.br/>. Accessed on: June 25, 2015.

FERRARI, Patricia Flôres. Environmental perception of hotel managers: a case study in Caxias do Sul (RS). Caxias do Sul: 2006.

FREDLINE, E. & Faulkner, B. Host **Communities Reactions: A Cluster Analysis**. Annals of Tourism Research, 27, (3), 763-784, 2005.

FREITAG, Thomas. Enclave tourism development: for whom the benefits roll? **Annals of Tourism Research**, v.21, n. 3, 1994.

FOURASTIÉ, Jean. **Leisure and tourism**. Rio de Janeiro: Salvat, 1979.

GAIA, Alexandre de Avila. **A method for managing environmental aspects and impacts**. Florianópolis: UFSC, 2001. Thesis (Doctorate) in Production Engineering - Federal University of Santa Catarina.

GARROD, B; FYALL, A. Beyond the rhetoric of sustainable tourism? **Tourism management**. United Kingdom: Elsevier Science, v. 19, n. 3, 1998.

GIESTA, Lilian Caporlingua. Sustainable development, corporate social responsibility and environmental education in the context of organizational innovation: concepts reviewed. **Revista adm. UFSM**, Santa Maria, v. 5, special edition, p. 767 - 784, dec 2013.

GODOI, C. K.; BANDEIRA-DE-MELLO, R.; SILVA, A. B. da (Org.). **Qualitative research in organizational studies:** paradigms, strategies and methods. Sao Paulo: Saraiva, 2006.

GONÇALVES, Luiz Claudio. **Environmental management in lodging establishments**. Sao Paulo: Aleph, 2004.

HARRINGTON, H. James; KNIGHT, Alan. **The importance of ISO 14000**: how to update your EMS effectively. Sao Paulo: Atlas, 2001.

IBGE - BRAZILIAN INSTITUTE OF GEOGRAPHY AND STATISTICS. **Census Demographic, 2014**. Available at: <www.ibge.gov.br>. Accessed on: July 10, 2015.

IGNARRA, Luiz Renato. **Basic concepts of tourism**. 2. ed. rev. ampl. Sao Paulo: Thomson, 2003.

INTERNATIONAL ORGANIZATION FOR STANDARDIZATION. **ISO 14001**. Survey.

ISOCentral Secretariat, Switzerland, 2012. Available at: <http://www.iso.org/iso/home.html>. Accessed on: June 20, 2015.

IVARS, J. A. **Planificación turistica de los espacios regionales**. Madrid: Sintesis, 2003.

IVANOV, Stanislav. "Tourism and Poverty", **International Journal of Contemporary Hospitality Management**, Vol. 24 Iss: 4, pp.674 - 676, 2012.

KO, T. G. "Development of a tourism sustainability assessment procedure: a conceptual approach", **Tourism Management**, Vol. 26 No. 3, p. 431 - 445. (2005).

KOHLRAUSCH, Aline K. **Environmental labeling то help form conscious consumers**. Dissertation (Master's in Production Engineering). Federal University of Santa Catarina - UFSC. Florianópolis, 2003.

KOROSSY, Nathâlia. From predatory tourism to sustainable tourism: a review of the origin and consolidation of the sustainability discourse in tourism. **Caderno virtual de turismo**. Rio de Janeiro, v. 8, n. 2, p. 1-13. 2008.

LANFANT, M.; GRABURN, N.H.H. International tourism reconsidered: the principle of the alternative. In: SMITH, V.L.; EADINGTON, W.R. (Eds). **Tourism alternatives**: potentials and problems in the development of tourism. Philadelphia: University of Pennsylvania Press and the International Academy for the Study of Tourism, 1992.

LEA, John. **Tourism and development in the third world**. London: Routledge, 1988.

LIU, A.; WALL, G. Planning tourism employment: a developing country perspective. **Tourism Management**, 27, pp. 159-70, 2006.

LUZ, C. A.; VIÉGAS, J. F.; FORNARI FILHO, P. Sustainability: an example of basic attitudes for managers to start practicing sustainable management in companies. **Revista Borges,** v. 04, n. 01,2014.

MALHOTRA, Naresh K. **Marketing research**: an applied orientation. 3. ed. Porto Alegre: Bookman, 2001.

MALTA, Maria Mancuello; MARIANI, Milton Augusto Pasquotto. Case study of sustainability applied to the management of hotels in Campo Grande, MS. **Revista Turismo Visao e Açao.** Mato Grosso do Sul, v. 15, n. 1, p. 112-129, jan-abr. 2013.

MARTINS, G. A.; THEÓPHILO, C. R. **Metodologia da investigaçâo cientifica para ciências sociais aplicadas.** Sao Paulo: Atlas, 2007.

MATTAR, Fauze Najib. **Marketing research**: compact edition. 5. ed. v.1, Sao Paulo: Editora Atlas, 1999.

MEBRATU, D. Sustainability and Sustainable Development: Historical and Conceptual Review. Environmental Impact Assessment Review, v. 18, p. 493 - 520, 19988.

MEDEIROS, L. C.; MORAES, P. E. S. Tourism and environmental sustainability: references for the development of sustainable tourism. **Environment and Sustainability Journal.** V. 3, n. 2, pp. 198-234, 2013.

MINISTRY OF TOURISM. National Tourism Plan 2013 - 2016. **"Tourism doing much more for Brazil".** Brasilia, 2014.

McCOOL, Stephen F.; MOISEY, Niel R. **Tourism, recreation, and sustainability**: linking culture and the environment. Wallingford: CAB International, 2001.

MOSSORÓ. **Tourism, 2015.** Available at: <www.prefeiturademossoro.com.br>. Accessed on: July 15, 2015.

MORAES, R. Content analysis. **Revista Educaçâo.** Porto Alegre, v. 22, n. 37, p. 7-32, 1999.

OLIVEIRA, O. J.; PINHEIRO, C. R.M. S. Implementation of ISO 14001 environmental management systems: a contribution from the people management area. **Gestâo da Produçâo**, v. 17, n. 1, p. 51-61, 2010.

UNWTO - World Tourism Organization. **Introduction to Tourism.** Sao Paulo: Roca, 2015.

PANATE, Manomaivibool. **Resources, Conservation & Recycling.** Vol. 103, p. 69 - 76. 8p. Oct, 2015.

PEARCE, Philip. **The relationship between residents and tourists**: research literature and management guidelines. In: THEOBALD, W. F. (Ed.). Global Tourism. Sao Paulo: Editora SENAC, 2001.

PETROCCHI, M.; Bona, A. **Agências de turismo**: planejamento e gestâo. 3. ed. Sao Paulo: Ed. Futura, 2003

PIRES, Fernanda. **Knowledge management applied to sustainable tourism management in national parks**. Thesis, 2010.

PNT - NATIONAL TOURISM PROGRAM. **Brazilian tourism GDP**. Available at: < http://www.turismo.gov.br>. Accessed on: Aug. 23, 2015.

PORTAL - PORTAL DA COSTA BRANCA. **History**. Available at: <www.portalcostabranca.com>. Accessed on: July 14, 2015.

SUSTAINABLE TOURISM CERTIFICATION PROGRAM - **NIH-54** - National standard for lodging facilities - Requirements for sustainability - Instituto de Hospitalidade, 2004.

ROSVADOSKI-DA-SILVA, P.; GAVA, R. E.; DEBOÇA, L. P. Economic structure and tourism: local versus extra-local dominance in the Lavras district of Ouro Preto (Minas Gerais, Brazil). **Journal of Tourism and Development**, 4(21/22), pp. 75-83, 2014.

RUDIO, Franz Victor. **Introduction to the research project**. Petrópolis: Vozes, 1999.

RUSCHMANN, Doris Van de Meene. **Sustainable tourism**: protecting the environment. São Paulo: Papirus, 1997.

. **Tourism and sustainable planning:** protecting the environment. Campinas: Papirus, 2008.

SANTOS, J. G.; CHAVES, J. L. A. Socio-environmental responsibility: a study of hotels in Gravatà- PE. In: XVI ENGEMA (International Meeting on Business Management and the Environment). São Paulo, 2014. **Proceedings...**

SEBRAE-MS. **Solid waste management:** an opportunity for municipal development and for micro and small enterprises -- São Paulo: Instituto Envolverde. Ruschel & Associados, 2012.

SOUSA, J.F.; FONSECA, C. C. **Projeto de Assistência Tècnica Juridica no Dominio da Reforma Portuària**, June 10, 2013.

SWARBROOKE, John. **Sustainable tourism management**. Wallingford: CAB International, 1999.

TACHIZAWA, Takeshy. **Environmental management and corporate social responsibility**. Sao Paulo: Atlas, 2008.

TRUNG, D.N.; KUMAR, S. **Resource use and waste management in Vietnam hotel industry**. J. Cleaner Prod., 13 (2005), pp. 109-116, Article, 2005.

TUNG, R.L. & AYCAN, Z. Key success factors and indigenous management practices in SMEs in emerging economies. **Journal of World Business**, 43, pp. 381-384, 2008.

VALLE, Cyro Eyer. **Environmental quality**: how to be competitive while protecting the environment: (how to prepare for ISO 14000 standards). Sao Paulo: Pioneira, 1995.

VARUM, Celeste Amorim; MELO Carla; ALVARENGA António; CARVALHO Paulo Soeiro de, (2011) "Scenarios and possible futures for hospitality and tourism", **Foresight,** Vol. 13 Iss: 1, pp.19 - 35.

VERGARA, S. C. **Métodos de pesquisa em administraçao**. 2. ed. Sao Paulo: Atlas, 2006.

WALPOLE, M. J.; GOODWIN, H. J. **Local economic impacts of dragon tourism in Indonesia Annals of Tourism Research**, v.27, n.3, 2000.

WHEELLER, Brian. **Tourism's troubled times**: responsible tourism is not the answer. Tourism Management, v.12, n.2, 1991.

WTO. World Tourism Organization. **Tourism**. Available at: <www.2.unwto.org>. Accessed on: June 25, 2015.

WORLD TRAVEL & TOURISM COUNCIL (WTTC) - **League**   **Table Summary**.

London, 2015.

WTTC, World Travel & Tourism Council. 2011. **The economic impact of travel and tourism.** Available at: <http://www.wttc.org/bin/pdf/original_pdf_file/world.pdf> Accessed on: June 25, 2015.

YIN, R. K. **Case Study:** Method Planning. Porto Alegre: Bookman, 2005.

ZAWISLAK, P. Antônio; GAMARRA, José E. T. The Importance of Specific Assets in the Differentiation of Firms in the Hotel Sector. **Revista Economia & Gestâo**, v. 15, p. 79-111, 2015.

7 APPENDICES

APPENDIX A - Informed consent form for the companies surveyed

POTIGUAR UNIVERSITY - UnP

GRADUATE PROGRAM IN ADMINISTRATION - PPGA

PROFESSIONAL MASTER'S IN ADMINISTRATION - MPA

Master's student: Francisco Tomaz Pacifico Jûnior

free and clear consent form

The Hotel is invited to participate in this

The general objective of this research is to identify the perception of hotel managers in Mossoró/RN in the face of a scenario of Corporate Environmental Responsibility and sustainability practices. The choice of topic arose from an interest in identifying the environmental management actions practiced by hotels, in view of this scenario of growing concern for the environment. The research will be carried out in two stages. In the first stage, the manager will be interviewed. The second stage will involve questionnaires with the hotel's employees. All the information collected, both in the questionnaire and in the interview, will be used only by the researcher in order to meet the objectives of the research and will be kept strictly confidential, thus ensuring the confidentiality and privacy of those who take part in the research. At no time, not even during the research, will the name of the hotel or the participants be mentioned. The data may be used during scientific meetings and debates and published, preserving the anonymity of the participants. By taking part in this research you will not receive any direct benefits. However, it is hoped that this research will lead to important reflections on environmental practices in hotel enterprises. Whenever the organization feels the need, it can ask for more information about the research by contacting the researcher and/or the educational institution to which he/she is linked.

I, _____ declare

that I am aware of the objectives and procedures of the research and, in a free and informed manner, as representative of the _____ CNPJ:

I express my interest in taking part in the research.

Signature of the person responsible for the organization under investigation

Signature of the researcher responsible

Mossoró, _____ 2016 _____.

APPENDIX B - INTERVIEW SCRIPT FOR MANAGERS

M A UP
Mestrado
ADMINISTRAÇÃO

POTIGUAR UNIVERSITY - UnP

POST-GRADUATE PROGRAM IN ADMINISTRATION - PPGA

PROFESSIONAL MASTER'S DEGREE IN ADMINISTRATION - MPA

Master's student: Francisco Tomaz Pacifico Júnior

Hotel identification

Name: _____

No. of rooms: No. of beds:

Annual occupancy rate:No. of employees:

What is the company's mission?_____

How many and which sectors are there in the hotel?

Questions

1. How can you describe the involvement of senior management in the process of implementing and maintaining environmental practices?

2. What is the main difference between this hotel and other hotel chains in Brazil?

3. Have you ever had the idea of adopting a positive attitude towards the environment? If so, why?

5. What are the main benefits of adopting environmental practices in hotels?

6. What were the main barriers to adopting environmental practices? _____

7. Are you aware of environmental legislation?

( ) Yes. Which laws? _____

( ) No.

8. Does your hotel do any kind of work to make guests and staff aware of environmental

issues?

If so, for how long and how often?

If not, why? _____

9. What resources are used to raise awareness?

( ) Posters ( ) *Websites* ( ) Leaflets ( ) Nature trails ( ) Other, which?

10. Do employees receive training on environmental issues? Which ones and how often?

11. Can you tell us where the water that supplies your hotel comes from?

( ) Municipal public service ( ) Artesian wells ( ) Other

12. Is the hotel's water potability standards being monitored?

If so, which one(s)? _____

If not, why? _____

13. Does your hotel use any methods to reduce water consumption?

If so, which one(s)? _____

If not, why? _____

14. Does your hotel use any methods to reduce energy consumption?

If so, which one(s)? _____

If not, why? _____

15. Is your hotel's sewage treated in any way?

( ) Yes, why? _____

( ) No, why? _____

16. Does your hotel separate solid waste?

( ) Yes, how? _____

( ) No. Why not? _____

17. Do you know the destination of the solid waste generated in your hotel?

( ) Yes, which one(s)? _____

( ) No.

18. Does your hotel take environmental factors into account when it outsources its services?

( ) Yes. What factors? _____

Why? _____

( ) No. Why not? _____

19. What skills, knowledge and attitudes are needed to take care of the environment?

I want morebooks!

Buy your books fast and straightforward online - at one of world's fastest growing online book stores! Environmentally sound due to Print-on-Demand technologies.

Buy your books online at
www.morebooks.shop

Kaufen Sie Ihre Bücher schnell und unkompliziert online – auf einer der am schnellsten wachsenden Buchhandelsplattformen weltweit! Dank Print-On-Demand umwelt- und ressourcenschonend produziert.

Bücher schneller online kaufen
www.morebooks.shop

info@omniscriptum.com
www.omniscriptum.com

Printed by Books on Demand GmbH, Norderstedt / Germany